Peter Ryan, a Fellow of the British Interplanetary Society and associate member of the American Institute of Aeronautics and Astronautics, studied Geology at Trinity College, Dublin. He spent four and a half years working for BBC TV during which time he was on the production teams of *Tomorrow's World* and later *24 Hours*. He became a freelance writer in May 1968, returning to the BBC as scientific adviser on the Apollo missions. Since Apollo 11 he has reported space missions for the Irish television and radio service. His work has taken him to space centres in Europe and the United States. In particular the Apollo programme has brought him into close contact with NASA and many specialists throughout the American industrial complex which builds and equips the spacecraft.

The Invasion of

Revised and Enlarged Edition

Peter Ryan

Penguin Books

the Moon

1957–70

Penguin Books Ltd, Harmondsworth,
Middlesex, England
Penguin Books Inc., 7110 Ambassador Road,
Baltimore, Maryland 21207, U.S.A.
Penguin Books Australia Ltd, Ringwood,
Victoria, Australia

First published as a Penguin Special 1969
Revised edition published in Pelican Books 1971
Copyright © Peter Ryan, 1969, 1971

Made and printed in Great Britain by
Hazell Watson & Viney Ltd, Aylesbury, Bucks
Set in Linotype Times

Contents

'That's one small step for a man, one giant leap for mankind.'
Neil A. Armstrong, Tranquillity base, 20 July 1969

'We were the first that ever burst
Into that silent sea.'
Samuel Taylor Coleridge. *The Rime of the Ancient Mariner*

Note: Armstrong placed his left foot on the moon at
10.56 p.m. Washington Time, 20 July 1969. In London the time
was 3.56 a.m., 21 July, and in Canberra it was 12.56 p.m., 21 July.

Preface to the Revised Edition

The first edition of this book was published as a Penguin Special on 8 August 1969, nineteen days after the splash-down of Apollo 11. At that time the mission was greeted as the exciting start to a rapid succession of lunar landings as well as the triumphant conclusion to a race, but in the following twelve months the second thoughts of American policy-makers considerably changed this perspective. It is now clear that phase one of man's exploration of space – the race to the moon – began with Sputnik 1, reached a climax with Apollo 11 and came to a definite end with a revealing coda in Apollos 12 and 13. In contrast, America's second phase of manned space exploration, beginning with Apollo 14, shows a new set of priorities: scientific pay-off is emphasized; cost is closely monitored; pace is much steadier.

Yet the first phase, and in particular the details of Apollo missions 11, 12 and 13, remains relevant to future lunar exploration as well as exciting in its own right. Neil Armstrong's trip to the Sea of Tranquillity will remain the classic lunar journey, retrospectively spotlit by Apollo 13's hint of what *could* have happened; so a detailed explanatory narrative of the nine-day mission of Apollo 11 remains the core of this book. An account of the Apollo 12 and 13 missions has been added for this edition and the review of future plans, both short and long term, has been considerably expanded. The revised edition has also provided an opportunity to improve on the photographs included in the first edition: they have been replaced by two new sets of plates, one in black and white, the other in colour.

Throughout the book certain conventions have been observed to make it clear who is talking at any given time. Conversations between the astronauts and Mission Control *have been set in italics,*

while commentaries during the mission from launch control at Cape Kennedy and Mission Control at Houston have been set down in smaller type.

Timings throughout missions are given in ground elapsed time or GET, that is, the time that had elapsed from the moment the Saturn V lifted off from its launch pad at Cape Kennedy at GET 00:00:00 (hours:minutes:seconds). For the more important phases of a mission GET has been converted to British standard time (BST). For readers in Australia the conversions are made as follows: BST plus 9 hours for Queensland, New South Wales, Canberra and Victoria; BST plus 8½ hours for Northern Territory and South Australia; BST plus 7 hours for Western Australia; for Tasmania BST plus 10 hours in the summer and BST plus 9 hours in the winter. For readers in New Zealand: BST plus 11 hours. For readers in the United States the conversions are: for EDT: BST less 5 hours; for EST and CDT: BST less 6 hours; for CST and MDT: BST less 7 hours; for MST and PDT: BST less 8 hours, and for PST: BST less 9 hours. For the Apollo 11 mission GET 00:00:00 was exactly 13:32:00.0724 Greenwich Mean Time, 16 July 1969.

I would like to thank the many men and women of the National Aeronautics and Space Administration, the Grumman Aircraft Engineering Corporation and the North American Rockwell Corporation whose advice I sought in writing this account. Without their help the book would have been impossible; nor would it have come into your hands without the advice and encouragement of many people at Penguin Books, Victor Hasselblad Aktiebolag and within the BBC. To these, thank you again.

'Their Germans are better than our Germans'

In the early hours of 4 October 1957 a two-stage rocket thundered its way through the scattered cloud high above Tyuratam near the Aral Sea in Soviet Central Asia. As it gathered speed, the four strap-on engines were discarded, these spent sections falling in a wide arc back to the ground. Finally the last engine shut down and the protective nose cone opened to reveal Sputnik 1, a twenty-three-inch, 184-pound sphere now circling the earth way above the atmosphere in the vacuum of outer space. Its speed of over 17,000 miles an hour was enough to balance the strong pull of earth's gravity.

In just ninety-six minutes the bleeping satellite completed its debut orbit. The Americans were stunned. That the space age dawned on a divided world was never more clearly demonstrated than by their first reactions. To them it was unacceptable that their technological prowess could ever let them down, let alone leave them in second place to a nation they had long written off as scientifically backward. At first, many Americans in high places simply refused to believe the Soviet announcements of the launch. Several members of Congress even dismissed Sputnik 1 as a hoax. To calm his nation President Eisenhower made a speech which must be one of history's best examples of lack of foresight at the top:

Now, as far as the satellite itself is concerned, that does not raise my apprehensions, not one iota. I see nothing at this moment, at this stage of development, that is significant in that development as far as security is concerned, except as I pointed out, it does very definitely prove the possession by the Russian scientists of a very powerful thrust in their rocketry, and that is important.

Later, in reply to a question, he added:

The value of that satellite going around the earth is still problematical, and you must remember the evolution that our people went through and the evolution that the others went through, from 1945, when the Russians captured all of the German scientists in Peenemünde.

The plea was to be that 'their Germans are better than our Germans' while the American administration thought what to do about it. But this was simply not true. The Peenemünde chief, Wernher von Braun, and the best of his team had surrendered by choice to the Americans and brought with them all their most important plans and records comprising several tons of documents. Besides, the Peenemünde Germans that had fallen to the Russians were not directly engaged on civil aspects of the Soviet space programme but were more concerned with developments of the V2 rocket for the Red Army.

Sputnik 1 was a purely Russian achievement and the man behind it, unknown even to the Russian public until his death in 1966, was Sergei Pavlovich Korolev. His story is very similar to that of von Braun in that both of them were members of small, fund-starved groups of the 1930s dedicated to space travel and both were to be responsible for two of the largest scientific programmes ever mounted. Korolev is known to have met the Russian father of space travel, Konstantin Tsiolkovsky, a man who was designing spacecraft before 1900 and who worked on the simple principle that 'the impossible today will become the possible tomorrow'. Tsiolkovsky is to Russian space history what the pre-World War One rocket-building pioneer Robert Goddard is to the Americans.

The Russians had chalked up an outstanding first, yet Sputnik 1 was more than an orbiting radio beacon. The launch came three months after the start of the International Geophysical Year (IGY), the result of sixty-four countries getting together for an intensive cooperative study of the planet earth taken as a whole. Early on, both Soviet and American scientists had proposed that an artificial satellite would be of great value to this ambitious programme, though nobody predicted that it would be the Russians who would achieve this during the IGY.

The first Sputnik provided information about the density of the upper atmosphere as it skimmed through these thin layers on its nearest approach each orbit. The radio signals it passed back through the ionosphere provided the first data necessary for the design of reliable communications with space vehicles. Above all, Sputnik 1 demonstrated that instrumentation could be developed to withstand the weightless vacuum and sudden temperature changes of space. All this was vital to the next phase of the Soviet space programme.

Meanwhile, America's international prestige was at stake. Naturally enough, the military establishment had their reasons too for getting into space, and getting there quickly, now that the Russians were in orbit.

At that time the American space effort was split up. It was not yet under a central body, as the Soviet programme was, and inter-service rivalries bedevilled it. First, the US Navy was deputed to launch a face-saving satellite. On 23 October 1957 a final test of their Vanguard three-stage rocket suggested that they were ready to go ahead. To reinforce the sense of urgency came a second Soviet launching, that of Sputnik 2 on 3 November 1957, containing the first living creature sent into space, the dog Laika. Using high altitude rockets, the Russians had already established that various living creatures could withstand both the stresses of take-off and landing and the limited periods of weightlessness afforded by these flights. Now they wanted to know the effects of protracted weightlessness in orbital flight. Clearly the Russians were already planning to put a man in space before the Americans had even launched a satellite.

The US Navy programme was rushed ahead in spite of the fact that its scientific directors were not happy, and on 6 December 1957, in front of a television audience of millions, America's first attempt to launch a satellite ended in a bright ball of orange flame on the launching pad.

The Navy having failed, the Army team was called in. This team consisted of von Braun and his Jupiter C rocket. Ironically, his proposal to use the Jupiter C to launch a satellite had been expressly turned down by the Department of Defense the year before. Von Braun went immediately to work and on 31 January 1958 Explorer 1, a thirty-one-pound orbiter, took the Stars and Stripes into the cosmos.

The instrument package aboard this first American satellite was the work of a team led by Dr William Pickering, who subsequently launched the project which brought close-up pictures back from Mars. Though smaller than Sputnik 1, the Explorer obtained more useful data than its heavier counterpart. Notably it led to the discovery of the Van Allen belts, hitherto undetected bands of radiation that surround the earth in two doughnut-ring shaped belts beyond the atmosphere.

At last, on 17 March 1958, after yet another technical hitch the US Navy managed to get their three-pound Vanguard 1 satellite into orbit. It was promptly christened 'The Grapefruit' by Krushchev. The next Russian launch was on 15 May: Sputnik 3 weighed nearly 3,000 pounds, of which two-thirds were instrumentation and power sources. Thus in the early days two facts stood out. One: there was clearly already a 'space race'. Two: the Russians seemed to most of the world to be winning it.

Target: the moon

In October 1958, to bring the space effort under one roof and to curb inter-service rivalry, the Americans founded the National Aeronautics and Space Administration. From the first, NASA was firmly committed to getting an American into orbit. By 1959 the first one-man Mercury space capsules were being designed and built and in April of that year the first seven astronauts had been chosen from among over 500 suitably qualified applicants.

Meanwhile, Russia was giving the sputnik programme a rest. There were no more Soviet orbital launches until May 1960, by which time America had put up eighteen earth satellites to Russia's three. The reason for the two-year gap between Sputniks 3 and 4 was that Russia had set her sights on the moon.

The Americans were the first to try reaching our only natural

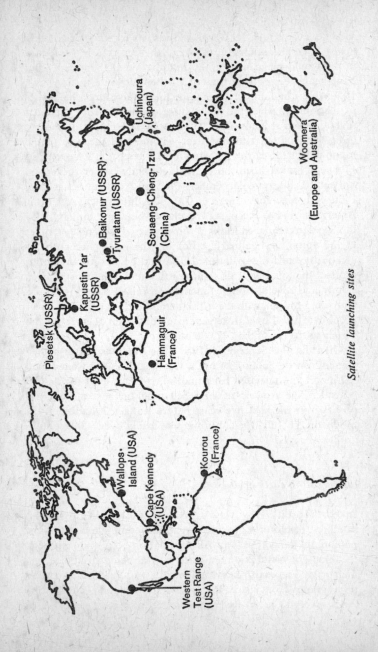

Satellite launching sites

satellite across quarter of a million miles of space. Quite apart from the scientific value of the mission, a successful probe would make up for the humiliation of Sputnik 1. But once again the rocketry was to let them down. On 17 August 1958 Able 1 had only managed the first twelve and a half miles of the journey when the lower stage of the launcher exploded seventy-seven seconds after launch. On 11 October, Pioneer 1 did a little better, reaching nearly a third of the way to the moon before the earth's gravity pulled it back. On 9 November, Pioneer 2 failed when the third stage of its carrier rocket did not ignite and, on 6 December, Pioneer 3 did no better than Pioneer 1, reaching a mere 66,000 miles into space. Again the Americans had to accept second place when on 2 January 1959 the Russians launched Lunik 1, which passed within 4,660 miles of the moon on 4 January, becoming the first space probe to escape the earth's gravitational field and go into its own orbit around the sun. Not all Americans could take this setback calmly and several Congressmen again raised the cry of hoax. Only on 3 March, with the fifth attempt, did America's Pioneer 4 emulate the Lunik 1 achievement but even then it got no nearer to the moon than 37,000 miles.

Then on 13 September 1959 the Russians crash-landed Lunik 2 on the moon and, on 4 October, Lunik 3 went round the moon and televised back man's first glimpse of the hidden far side. The Americans launched nine further unsuccessful probes over the next five years before Ranger 7 landed on the moon on 31 July 1964, having sent back over 4,000 photographs.

Spacecraft in earth orbit

In earth orbit, 1959 was a year of exclusively American satellites. Explorer 6 provided an almost complete chart of the potentially dangerous Van Allen radiation belts and sent back the first TV pictures of the earth before being disabled by a meteorite. Vanguard 2 transmitted the first primitive weather photographs from space. It was also the year the United States Air Force was given its chance to get into space. The USAF

Discoverer programme was a military programme and from 1962 it continued only in secret. Its beginnings were rooted in the need to replace the now vulnerable U2 spy planes, and the USAF was interested in developing satellites for, among other things, photographic reconnaissance, missile launch detection and navigation. But the Discoverer series of satellites achieved a number of important 'firsts' also relevant to the project of sending man into space. These included stabilizing a satellite in orbit, changing the orientation of a satellite in its orbit, changing an orbit, ejecting a capsule from a satellite, and recovering the capsule from orbit. In particular, the success of the ejection and recovery process was as useful to the civilian developers of the Mercury spacecraft that were to carry first monkeys and then men as it was to the USAF in collecting their spy films.

On 9 September 1959 an empty test version of the Mercury capsule made its first sub-orbital hop, while on 28 May 1959 a pair of monkeys, Able and Baker, survived a 300-mile-high flight inside the re-entry cone of one of the Army's Jupiter ballistic missiles.

In 1960 the Americans launched nine scientific payloads into earth orbit. These included the world's first weather satellite, Tiros 1, which transmitted nearly 23,000 cloud cover pictures between April and June, and the first passive communications satellite, Echo 1, a giant silvery balloon. Once inflated in space, it reflected speech and television signals back to the earth. By October the first active communications satellite, Courier 1B, was in orbit. Earlier, in August, the first Discoverer capsule had been safely recovered from the sea.

Another recovery from earth orbit was also successfully carried out that month, that of the re-entry section of Sputnik 5. This cabin contained the Russian dogs, Belka and Strelka, who had completed eighteen orbits. In May the similar but empty cabin of Sputnik 4 was not retrieved when it went into a higher orbit instead of returning to earth. This test had followed the apparently successful 700-mile-high space hop of the nose cone of a ballistic missile recovered 7,760 miles down-range in the Central Pacific. Sputniks 4 and 5 were prototypes of the Vostok spacecraft designed to carry a man.

Launched in December, Sputnik 6 or Spacecraft 3 also

15

carried two dogs, Pshcehelka and Mushka, but unhappily they came to a fiery end when the automatic stabilization of the craft failed, causing too steep a re-entry path.

1960 was the final year of preparation for man's entry into space in person. While the Russians were flying their Vostok prototypes, the Americans were testing the Mercury capsule. On 29 July 1960 the second unmanned Mercury craft sat atop an Atlas rocket, originally developed to carry nuclear warheads, at what was then called Cape Canaveral (now Cape Kennedy). A minute after launch the USAF-developed rocket exploded. Once again von Braun came to the rescue, and his smaller Redstone rocket was pressed into service. By 21 November the first Redstone was ready on the pad. The button was pressed and the rocket lifted the first few inches. Then it shut down. Fortunately it did not explode and the capsule was undamaged. In December another Redstone behaved perfectly.

A crucial date in 1960 was 2 July when the American Government's Scientific and Space Committee proposed a manned expedition to the moon before 1970. On 29 July, NASA announced officially that a programme called Apollo would follow the Mercury series. Already von Braun was at work on developments of his successful Jupiter and Redstone rockets: 'With this go ahead we began design work on a missile and space booster that could be built from a maximum of proven components with the least possible expenditure of time, money and effort.' Thus was the Saturn series of rockets born.

That year also saw the first, though unsuccessful, attempts to open up a third avenue of space exploration, the path to the planets. The Americans failed to get Pioneer 5 anywhere near its target, Venus, while Russia made two unannounced attempts to send a probe to Mars.

Meanwhile, twelve Soviet cosmonauts including two Soviet Air Force pilots, Yuri Gagarin and Herman Titov, were undergoing intensive training, as were the seven American astronauts. The composition of this first American team – three Navy, two Air Force and two Marine Corps pilots – helped to minimize inter-service jealousy. Their names: Scott Carpenter, Gordon Cooper, John Glenn, Virgil Grissom, Alan Shepard,

Walter Schirra and Donald Slayton. 1961 was to see two cosmonauts and two astronauts in space.

The first spaceman

January 1961 saw the dawn of the Kennedy–Krushchev era. Whatever this was to mean to others, to both American and Soviet space programmes it meant the age of the blank cheque. In 1961 NASA spent one billion (1,000,000,000) dollars, over half the total national space budget, as the race to put the first man into orbit approached the finishing line.

On 31 January a chimpanzee called Ham made a 420-mile trip in a Mercury capsule propelled by a Redstone rocket, and in spite of a rough landing was recovered safely, but the Redstone was limited to such sub-orbital hops. To orbit the Mercury capsule, the problems with the more powerful Atlas had to be cleared up. On 21 February American hopes were raised when an Atlas took the craft on a 1,425-mile hop, re-entering the atmosphere at some 13,000 miles an hour.

In the first two months of the year there were few signs of Russian activity except for two more or less unsuccessful shots at putting a planetary probe en route for Venus. Then on 9 March a dog by name of Chernushka completed just one orbit aboard Sputnik 9 (Spaceship 4). On 25 March another dog, Zveozdochka, repeated the mission with Sputnik 10 (Spaceship 5). The Russians were now ready, the Americans were not.

At that time Western correspondents in Moscow were understandably getting a little trigger-happy. Faced with Soviet secrecy, every rumour hit the headlines. Thus on 11 April a story came out of Moscow about Vladimir Ilyushin, the aircraft designer's son, who had, according to a certain Western press agency, been into space on 7 April. Again understandably, the Russians made absolutely no comment.

A day later the real thing happened.

At 5.30 a.m. Moscow time on 12 April 1961 Yuri Gagarin was woken by the cosmonauts' doctor, as was his back-up pilot, Herman Titov. His breakfast was chopped meat, blackberry jam and coffee. He was helped into his spacesuit and

17

set off by bus to the awaiting Vostok (East) spacecraft at the Baikonur cosmodrome to the north-east of Sputnik 1's launch pad at Tyuratam. By 7.30 a.m. he was strapped inside the space cabin on top of the 125-foot, three-stage rocket. The cabin, he later said, smelt of spring fields.

Then at seven minutes past nine Moscow time the first man to enter the weightless void of space began his noisy journey. 'Off we go, everything is normal,' he reported as the Vostok booster inched off the pad. 'How magnificent, I see the earth, forests, clouds,' he said as the spaceship climbed higher. At over 17,000 miles an hour he journeyed over southern Russia and India to loop around south of Australia and out across the Pacific. Then, by way of Cape Horn and the southern half of Africa, the Vostok reached the point where it automatically re-aligned itself for its plunge back to earth. One hundred and eight minutes after lift-off the retro-rockets fired, and soon Gagarin was watching the increasing glow as the outer layers of the spacecraft's protective shield began to burn up on re-entering the earth's atmospheric envelope. Weightlessness ceased as the craft braked, and soon he experienced the strong G forces of deceleration. At 10.55 a.m. the charred capsule landed by parachute near a village called Smelovaka near Saratov, some six miles from the planned landing area. His arrival from space was witnessed by a woman, a girl and a cow. The woman, the wife of a forester, asked the orange spacesuited apparition whether he had really come from outer space. 'Just imagine it, I certainly have,' replied Gagarin.

The moon race begins

While this conversation was taking place, a disappointed but determined team at Cape Canaveral digested the Tass news agency bulletins. 'Let the capitalist countries try to catch up,' crowed Krushchev. Kennedy's reply came on 25 May:

I believe that this nation should commit itself to achieving the goal, before this decade is out, of landing a man on the moon and returning him safely to earth. No single project in this period will

18

be more impressive to mankind, or more important for the long-range exploration of space; and none will be so difficult or expensive to accomplish.

The gauntlet was down and the race to the moon was on.

Exactly one month earlier the Americans had suffered yet another setback with their Atlas booster. Meant to put an unmanned Mercury capsule into orbit, it veered off course and the range safety officer was forced to reach for his destruct button. The first American in space would have to be content with a space hop aboard the reliable Redstone.

For over four hours in the early hours of 5 May 1961, Alan B. Shepard waited in his Mercury capsule codenamed Freedom 7 while technical holds and clouds halted the countdown. Finally, at 9.24 a.m. Eastern Standard Time, the Redstone ignited. Forty-two seconds after lift-off it shut down as planned while Shepard hurtled 196,000 feet above the Atlantic at 4,500 miles an hour. As the Mercury capsule climbed to its peak altitude of 116 miles, Shepard manually adjusted its attitude. He was able to bring the craft to within five degrees of the required re-entry attitude before handing it back to automatic control. Fifteen minutes and twenty-two seconds after lift-off he was pitching quietly in the Atlantic.

On 21 July 1961 another of von Braun's Redstones carried another Mercury capsule, Liberty Bell 7, with Virgil 'Gus' Grissom aboard. Like Shepard, he 'flew' his craft manually during the space hop. Back in the Atlantic, Grissom had to swim for it when the capsule began to sink. A rescue helicopter soon picked him up but the weight of the waterlogged Liberty Bell proved too heavy for another helicopter and sank. It was nearly a costly disaster, but Grissom suffered no ill effects.

The next step to get an American in orbit required yet more tests of the troublesome Atlas. Before these were complete a second Russian, Gagarin's back-up pilot Herman Titov, was in orbit. Launched on 6 August, he completed seventeen orbits with Vostok 2, a manned repeat of the Belka and Strelka flight aboard Sputnik 5. It was not all plain sailing, for Titov suffered from a complaint that was to become familiar to astronauts: space sickness, similar to seasickness and caused by disorientation of the inner ear.

By 13 September the Americans had at last successfully orbited an empty Mercury capsule with the Atlas booster. On 29 November a chimpanzee by name of Enos made a two-orbit flight. Atlas was now ready: an American could go into orbit.

The year ended with no Soviet scientific satellites apart from two sputniks, now thought to have been the two halves of an unlucky Venus probe. America meanwhile had added fourteen unmanned orbiters, among them the first spy satellite and two miniaturized nuclear reactors. She failed in two attempts to crash-land a Ranger television probe on the moon.

On 20 February 1962 John Glenn, a US Marine Corps pilot, became the first American in orbit. In Mercury capsule Friendship 7, he completed three orbits in a flight that was far from trouble-free. An attitude control nozzle became clogged and he had to override the automatic system and fly the Mercury craft himself. On the final orbit telemetry signals indicated to ground controllers that the capsule heat shield might be loose. Fortunately this was not the case, for the results would have been fatal.

On 24 May Mercury pilot Scott Carpenter completed three orbits aboard Aurora 7. He showed that astronauts can be quite human by using much of the attitude control system's limited fuel supply getting in position to photograph sunrises. However, this did not go down very well with NASA. At the end of the third orbit the craft's retro-rocket engines failed to ignite and he had to start the descent manually by pushing a button. As a result the retro-burn started four seconds later than planned and Carpenter landed 250 miles from the planned point in the Atlantic.

Next it was again Russia's turn. Two Vostoks, 3 and 4, were launched on 11 and 12 August. Andrian Nikolayev aboard 3 and Pavel Popovich piloting 4 came to within three miles of each other. Nikolayev completed a record sixty-four orbits, nearly four days in space, while Popovich came down after forty-eight revolutions. During these Vostok flights two television cameras showed Nikolayev to Russian audiences below.

Meanwhile the Americans were ironing out the troubles with the Mercury capsule and on 3 October 1962 Walter Schirra

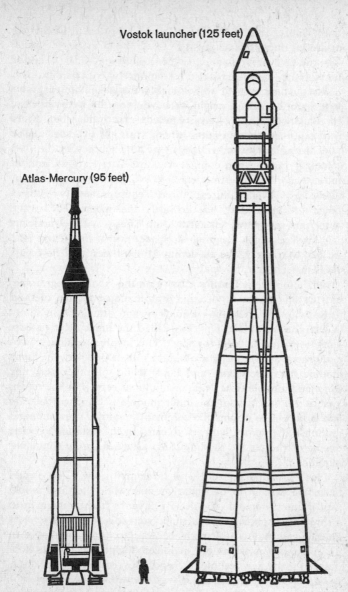

Vostok launcher (125 feet)

Atlas-Mercury (95 feet)

Mercury and Vostok launch vehicles

took Sigma 7 on a three-orbit trip which proved to be far less hazardous than previous flights.

In the field of unmanned space exploration America added the television-bouncing Telstar 1 among twelve scientific payloads launched in 1962 while earth-orbiting satellites of more or less secret military origin were being put up by both sides. Three more American Rangers failed to reach the moon. Mars also came in for attention when Mars 1, a Soviet probe launched on 1 November, began its long journey to the Red Planet. It failed when contact was lost after sixty-six million miles. According to American records there were other unannounced Soviet failures: three Venus probes, two others aimed at Mars and one towards the moon. These, the Americans say, never left earth orbit. That year the Americans had two shots at sending a space probe to Venus; the second, Mariner 2, got to within 21,000 miles as it flew past the planet.

1962 also saw increasing activity on the Apollo programme. In July an important decision was made about the method to be adopted for landing a man on and bringing him safely back from the moon. This was called the lunar orbit rendezvous technique or LOR for short. This involves sending a two-part spacecraft into orbit around the moon. One part, the lunar module (LM), separates and lands while the other part, the command module (CM), remains in lunar orbit. For the return journey the top half of the lunar module is launched back up into orbit where it joins the command module. The command module is then propelled back to earth by the restartable rocket engine of the service module (SM) attached to the command module.

This technique called for a three-man crew. Two would make the descent in the lunar module while the third would wait in the command module. To advance from a one-man to a three-man spacecraft in a single step was for many reasons impractical. A lot of experience was necessary, not only in multi-manned missions but in longer flights and the development of many new techniques, especially rendezvous and docking manoeuvres. Therefore it was decided in December 1961 to follow the one-man Mercury missions with the two-man

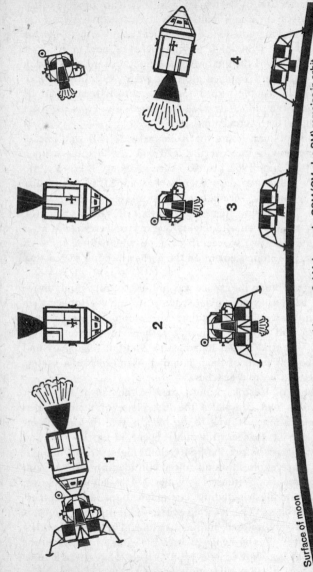

Surface of moon

1 SM engine puts craft into moon orbit   2 LM descends, CSM (CM plus SM) remains in orbit

3 Ascent stage of LM launched to rendezvous with CSM   4 LM discarded, SM engine starts journey home

Lunar orbit rendezvous (LOR) technique adopted for Apollo mission

Gemini series and in September a further team of nine astronauts was chosen which included Neil Armstrong. Then aged thirty-two and a graduate in aeronautical engineering, he was a NASA test pilot flying high altitude rocket planes.

On 15 May 1963 the final Mercury flight took place. Gordon Cooper with Faith 7 stayed aloft for twenty-two orbits. Once again the old trouble with automatic controls meant that he had to steer himself safely back but otherwise it was the most successful of the Mercury missions.

In terms of man hours the Russians were still way ahead. They were now to improve this lead and to add some woman hours. On 14 June 1963 Vostok 5 carried Valery Bikovsky into the cosmos, and two days later he was joined by twenty-six-year-old Valentina Tereshkova in Vostok 6, the first woman in space. The mission was similar to that of Vostoks 3 and 4. Bikovsky came down after eighty-one orbits, Tereshkova after forty-eight. The two spacecraft came to within three miles of each other, no improvement on the earlier meeting of Vostoks 3 and 4.

Putting a woman into space was not just a propaganda move as some have asserted. Soviet space plans include longer-term objectives than a moon landing. Space colonies would need both men and women. It is not perhaps too far-fetched to think of Valentina's later marriage to Andrian Nikolayev and the subsequent birth of their daughter Alenushka as a unique experiment in cosmic genetics.

The flight of Vostoks 5 and 6 marked the end of the series; the Russians had completed the first phase of their manned space programme. It was to be over a year before a new manned Soviet spacecraft would appear in earth orbit and nearly two years before the first Gemini flight. Meanwhile unmanned space exploration advanced considerably. Apart from the Department of Defense's activities, NASA launched sixteen scientific satellites including communications and weather-watching orbiters. The Russian score was thirty-five, many of them of an undisclosed military nature. There was another Soviet shot at Venus while in July 1964 the American probe Ranger 7 at last showed television viewers what it was like to hit the moon at 5,000 miles an hour.

24

In October 1963 NASA announced the names of a third group of trainee astronauts. They included Edwin Aldrin and Michael Collins. Aldrin, then aged thirty-three, had been a pilot in the Air Force when he joined NASA. In 1963 he was completing a second degree in astronautics. Collins, also aged thirty-three and in the Air Force, was a test pilot with a science degree.

By the end of 1964 development and testing of the various components of the Apollo spacecraft was well in hand and at Cape Kennedy a billion-dollar project was turning swampy Merritt Island into a moonport.

Once more a surprise from the Russians : they too had been working on a three-man space cabin. On 12 October 1964 it was aloft. Its crew consisted of Vladimir Komarov and two civilian cosmonauts, Boris Yegorov, who was a doctor, and a scientist, Konstantin Feoktistov. The weight of their Voskhod (Sunrise) was 11,731 pounds which is only about 300 pounds lighter than the Apollo command module. Like the Apollo capsule, the Voskhod design permitted the cosmonauts to dispense with their cumbersome spacesuits and enjoy a 'shirt-sleeve' environment.

On 18 March 1965 came a second Voskhod flight with Pavel Belyayev and Alexei Leonov which achieved another space 'first'. Opening the spacecraft hatch, Leonov spent twenty-four minutes as a one-man satellite floating in the vacuum of space with just his spacesuit protecting him. Man had taken his first walk in space, an airless vacuum in which his blood would soon have boiled had he ventured out unprotected. On their first attempt to return, the automatic re-entry system failed and the Voskhod 2 pilots had to bring the spacecraft down themselves an orbit later. The resultant change in landing site meant a long wait in the cold winter before rescue helicopters could pick them up. Leonov and Belyayev were the last Soviet cosmonauts in orbit for over two years. In that time the Americans were to complete their entire Gemini programme, surpassing the Russian record in both duration of flight and experience of different types of manoeuvre.

Following two unmanned flights of the Gemini capsule early in 1964, 'Gus' Grissom, veteran of the Mercury series, and John Young orbited three times on the Gemini 3 mission, launched on 23 March. For the first time a manned craft was moved from one orbit to another, a technique essential to the rendez-vous in space envisaged by the moon landing plans.

On 3 June, James McDivitt and Edward White flew sixty-two orbits aboard Gemini 4, the first really long-duration American trip. It included a twenty-one-minute space walk by White during which he manoeuvred himself about using gas jets from a small space-gun.

On 14 July the Americans added another feather to their space hat when, from 134 million miles across space, Mariner 4 sent back television pictures of Mars. The close-up frames showed an arid-looking planet with craters similar to the moon's. There were no signs either of the famous canals or of any Martian life. Launched on 16 November, Russia's Venus 3 later landed on that planet but neither it nor subsequent Soviet probes survived on the Venusian surface long enough to tell us very much about that cloud-covered inferno.

Gordon Cooper, another Mercury veteran, and Charles Conrad, a newcomer, were chosen for Gemini 5 which flew for a full eight days, the duration of an Apollo moon trip. They began their 120 orbits on 21 August 1965. Trouble with fuel cells which provided the spacecraft with electrical power nearly brought the mission to an early conclusion and some experiments had to be abandoned. However, the fact that NASA had now shown that it could build the spacecraft and train the men to fly in space long enough for a moon trip offered much hope for the future.

With Gemini 6 came the first American attempt at rendez-vous. A target vehicle was to be put into orbit and on its second revolution the Gemini craft was to follow it. On 25 October an Agena target was launched but a fault destroyed it and the mission was cancelled. Walter Schirra and Thomas Stafford climbed out of their capsule. A long delay seemed inevitable before a new target could be got ready for launch but this problem was solved by a radical change in plan: NASA decided to delay Gemini 6 until Gemini 7 was ready; the rendezvous was now to be between the two manned craft.

During this pause the French joined the ranks of the space powers. On 26 November 1965 a French-made Diamant rocket put an eighty-eight-pound, black-and-white-striped sphere into orbit from a launching pad in the Algerian Sahara.

James Lovell and Frank Borman in Gemini 7 were launched from Cape Kennedy on 4 December. (In the course of a flight lasting until 18 December they were to complete a record number of earth orbits – 206.) On 12 December, Schirra and Stafford climbed into their Gemini 6 capsule for the second time. Everything seemed fine and then, seconds after ignition, the motors of the Titan carrier rocket shut down. Had Schirra and Stafford used the emergency ejection seats – standard procedure in this situation – launch would have been delayed until well after Gemini 7 had returned. In the event, they elected to hang on and their courage enabled them to get off the ground on 15 December, their third attempt to get into orbit. For over five hours on that day Gemini 6 and Gemini 7 flew in formation, coming to within a few feet of each other.

Following that success, the next flight was to be an attempt at linking two spacecraft in orbit. This docking manoeuvre, vital to the moon landing plan, was to be the highlight of the Gemini 8 mission. On 16 March 1966 both the Agena target vehicle and the manned capsule were up in orbit. Piloting Gemini 8 were Neil Armstrong and David Scott. Locating and docking with the Agena presented no difficulty, but twenty minutes later trouble started. One of the stabilizing jets on the Gemini capsule failed to turn off and soon the two linked craft were in a wild spin. 'We've got serious problems here. We're tumbling end over end,' Armstrong reported but he managed to separate from the Agena and get the faulty thruster switched off. To regain control of the Gemini capsule had used up three-quarters of the re-entry fuel supply. Obviously the mission could not go on for the intended three days and Gemini 8 was brought back safely to an emergency splashdown after less than eleven hours aloft. Armstrong's coolness earned him much respect.

On 18 May 1966 another Agena target vehicle was launched but a failure in the Atlas booster deprived it of a position in orbit. By now NASA were prepared for such setbacks and on 1 June a substitute was in orbit. On 3 June, Thomas Stafford and Eugene Cernan followed in Gemini 9. Three orbits later they met up with the target vehicle but here they encountered a hitch. The nose fairing of their docking partner had not completely detached and the docking collar was still covered. A link-up was impossible and all they could do was to photograph the target which they nicknamed the 'angry alligator'. The mission went on for the planned forty-five orbits and Cernan made a partially successful space-walk, though an attempt to test an astronaut manoeuvring unit, a kind of mobile space armchair, had to be given up when his visor misted up.

On 18 July 1966 Gemini 10 with John Young and Michael Collins followed a further Agena target into orbit. Once docked, they started up the Agena's engine and propelled the two linked spacecraft up into a record 474-mile-high orbit, just below the potentially dangerous Van Allen belts discovered by America's first satellite in 1958. In February and March two Russian dogs, Veterok and Ugolyok, had spent twenty-two days in space, their 330 orbits taking them repeatedly through the Van Allen layers. On their return to earth it had been found that their muscles had grown limp and that their bones had lost calcium. It was some five days before they were back to normal. Yet the fact that they had survived this long ordeal indicated that a brief passage through these radiation layers such as would occur on a trip to the moon and back would not constitute a serious hazard.

Leaving the Agena 10, the astronauts then went off in search of the Agena 8 target vehicle with which Armstrong had so much trouble. Once they had found it Collins floated out into space to retrieve a micrometeorite detector and replace it with a fresh one for later collection by the Gemini 12 crew. Unfortunately Collins let go of the detector as well as a Hasselblad camera. Incidentally should anyone find the camera a reward of $10,000 awaits him.

The last two Gemini missions were to start in an atmosphere of elated confidence. Many of the major manoeuvres necessary to the Apollo mission had been tried with success. One vital test remained. When the returning astronauts leave the moon's surface in the top half of their lunar module, their rendezvous with the command module in orbit around the moon has to be completed within the first revolution. This needed to be rehearsed. On 12 September 1966 an Agena docking target was met to within half a second of the appointed time by the Gemini 11 capsule carrying Charles Conrad and Richard Gordon and docking was achieved within two-thirds of an orbit. The flight of Gemini 11 went on for nearly three days while the astronauts practised the rendezvous procedure several more times. During the second day of the flight Gordon took a space walk and attached a rope to the Agena target vehicle.

Gemini 12, though troubled with a series of minor faults, gave James Lovell and Edwin Aldrin a chance to practise link-ups in space. The failure of the main engine of the Agena deprived Aldrin of a chance to become the highest pedestrian in the world: he had to be content with three space-walks at lower altitudes. The flight lasted fifty-nine orbits and marked the end of the Gemini series. The launch pads at Cape Kennedy were now cleared for the moon missions.

By the end of 1966 the unmanned exploration of the moon had been stepped up by both the Americans and the Russians. In January the Soviet probe Luna 9 had landed safely and sent back the first pictures of the crumbly moonscape at close range. America's Surveyor I followed suit in June. In March, Russia had given the moon her first artificial satellite, Luna 10, and in August the Americans launched the first of their Lunar Orbiter series. These camera-carrying Lunar orbiters enabled highly accurate maps of the moon's surface to be prepared prior to the selection of future landing sites.

Setbacks

1967 was a year of disaster for both Soviet and American space programmes. In January three astronauts, Gus Grissom, Roger Chaffee and Edward White, died in a fire which broke out in an Apollo capsule on the launch pad at Cape Kennedy. In April,

29

Vladimir Komarov died when his Soyuz (Union) 1 spacecraft crashed after eighteen orbits. These accidents held up manned space exploration for over a year and a half.

The fire at Cape Kennedy was caused by an electrical fault which sparked the pure oxygen atmosphere of the Apollo cabin into a blazing inferno. The astronauts, fully spacesuited, were incinerated within seconds. There was no time to open the complicated hatch. After an exhaustive inquiry which went on for several months the Apollo capsule was radically rebuilt with fireproof materials, and a completely new, easily opened hatch was designed and fitted.

Komarov's death was caused by the failure of the Soyuz spacecraft's parachute system. It was a bitter blow in the year of the fiftieth anniversary of the revolution. On 27 March 1968 came another tragedy. Yuri Gagarin was killed when his plane crashed on a training flight.

Saturn and Apollo

The Apollo fire had come during a rehearsal of the first manned launch of this new spacecraft planned for 21 February 1967. Already three successful unmanned tests of the capsule which was to carry the first Americans to the moon and back had been completed. This first manned flight was now not made until 11 October 1968 with the Apollo 7 mission when a two-stage Saturn 1B launcher (a smaller member of the Saturn family which had now grown to include the Saturn V moon rocket first flown on 8 November 1967 in an unmanned test) lifted Walter Schirra, Donn Eisele and Walter Cunningham at the start of eleven days in orbit.

There they practised 'transposition and docking', a manoeuvre with the second stage of the Saturn 1B which soon would be made with the lunar module, first in earth orbit and then on the way to the moon. This involved separating the command service module (CSM) from the Saturn final stage, turning the CSM round and then docking back with the Saturn which was simulating the lunar module on the Apollo 7 flight.

During the mission the crew sent back the first of what were

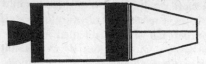

1 CSM separates from last stage of Saturn rocket

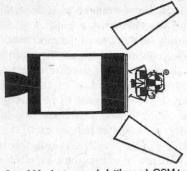

2 LM adapter panels jettisoned, CSM turns round

3 CSM approaches LM

 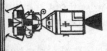

4 CSM docks with LM

 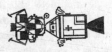

5 CSM and LM separate from last stage of Saturn rocket

Transposition and docking manoeuvre

to become regular television programmes from the Apollo capsule. At first Schirra, the commander of Apollo 7, was against this additional burden but later he relented and gave his audience some remarkable demonstrations of weightlessness. A toothbrush floating nonchalantly in mid air was included in the first performance. The Apollo 7 mission was concluded: NASA had proved the new command module. Only the lunar module now remained to be tested in space.

The lunar module must surely rank as one of the most extraordinary looking vehicles ever conceived and built by man; a four-legged craft designed for flight in the vacuum of space and above the airless moon, it is quite unlike anything that has been constructed before. It does not need to be aerodynamically smooth and it is meant to land only under the influence of the moon's gravity, one-sixth of that on earth.

To design and test this vehicle in the gravity and atmosphere of the earth was clearly not possible, so to gain some experience on earth, curious flying machines similar in appearance to 'flying bedsteads' were constructed and future lunar module pilots learned to fly them. They proved troublesome and dangerous. Concern mounted as crashes and narrow escapes gave rise to more and more modifications. Following the fire in January 1967, the LM was rebuilt from less flammable materials. Weight and stress problems too had to be contended with.

While NASA combated the delays, they went ahead with the next Apollo mission without the lunar module; Apollo 8 was to fly three men ten times round the moon and then return to earth.

On 26 October 1968 the Russians made their comeback. Georgy Beregovoy aboard Soyuz 3 was up in orbit. He too showed television audiences round his two-roomed spacecraft, which was obviously able to accommodate more astronauts. Staying up for sixty-four orbits, he came within 200 yards of the empty Soyuz 2 launched on 25 October. Docking did not take place but it was clear that the Russians were now ready to try this manoeuvre with manned craft. In October 1967 they had linked two unmanned Cosmos satellites. Whether this was a rehearsal for a space station made up of many such link-ups of separate units or part of their moon programme remained a subject of discussion in the West.

The plan to fly Apollo 8 round the moon at Christmas was not originally part of the American programme. It was partly an insurance against further delay in the Apollo flights, due to the troublesome lunar module, and the bad publicity this might attract. Certainly Apollo 8 was to be in keeping with part of President Kennedy's speech in 1961: 'No single space project in this period will be more impressive to mankind . . .'

On 21 December 1968 at exactly the appointed minute Frank Borman, James Lovell and William Anders were punched into earth orbit by the massive 363-foot Saturn V rocket. It was the first time that the giant rocket had carried human passengers and only the third time it had ever flown. After one revolution in earth parking orbit the Apollo 8 crew restarted the rocket motor of the third stage of Saturn V to increase their speed from 17,400 to 24,200 miles an hour. At this speed they began to pull away from the influence of earth's gravity and head out for the moon a quarter of a million miles away. 24,200 miles an hour is just below the 'escape velocity' at which a spacecraft would leave the planet for ever and pass into its own orbit around the sun. At a slightly lower velocity the Apollo 8 craft was now on a course that would carry it round the back of the moon. A combination of the moon's own gravity and the earth's would bring it back to earth without any further correction. A decision on whether to go into lunar orbit could be taken later.

By the afternoon of Christmas Eve they reached their destination and swinging behind the moon, out of sight from the earth, they started up the motor of the service module to slow themselves down into a sixty-nine-mile-high lunar orbit. It was at this point that the astronauts saw the moon in close-up for the first time. Frank Borman remarked on its desolation and hostility but conceded that it had an awesome beauty. The moon, it turned out, was a comparatively colourless place, sometimes a greyish white, sometimes a sandy brown. The absence of any atmosphere made the sunlight harsh, the shadows deep and dark. On Christmas Day they transmitted

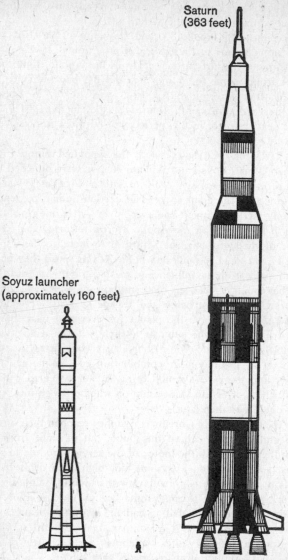

Saturn
(363 feet)

Soyuz launcher
(approximately 160 feet)

Soyuz and Apollo launch vehicles

pictures of the airless lunar landscape while they read aloud the beginning of the Bible.

The crew were deeply moved by what they were seeing but none the less their primary concern was the proper functioning of the spacecraft systems in deep space. Frank Borman had been the first man to enter the burned out capsule following the fire in January 1967 and above all it was his confidence in the new Apollo craft that mattered. He was, after all, the commander of a team of test pilots. Everything went well and after ten orbits round the moon they were ready to come home. The critical firing of the service module engine on Christmas night began their homeward path.

They jettisoned the service module and re-entered the earth's atmosphere on 27 December at a speed of over 24,000 miles an hour. Splashdown in the Pacific some ten minutes later ended man's first journey to the moon. America could fly its astronauts round the moon but they had yet to fly the craft that would land them on its surface. Apollo 9 was to be the first of only two test flights of the lunar module before the first moon landing would be tried. The date for that historic mission was now provisionally fixed for the summer.

Another manned mission widely predicted in December took place on 14 January 1969 when Russia launched Soyuz 4 with Vladimir Shatalov aboard into orbit. The following day Soyuz 5, carrying three cosmonauts – Boris Volynov, Alexei Yeliseyev and Yevgeny Khrunov – joined Shatalov. On the morning of 16 January the two spacecraft began an automatic approach and then, controlling his ship manually, Shatalov docked with Soyuz 5. The operation was televised. Next Khrunov and Yeliseyev put on their spacesuits, transferred themselves outside and entered the cabin of Soyuz 4, leaving Volynov in Soyuz 5. The two spacecraft then separated and the next day Soyuz 4 landed, leaving Volynov in orbit. He returned to earth on 18 January.

The flight of Soyuz 4 and 5 demonstrated the Soviet capability of assembling manned units in space from separately launched components. The concept of constructing space stations in orbit is a stated part of the Russian space programme but how this fits in with their moon plans is uncertain. Whether

35

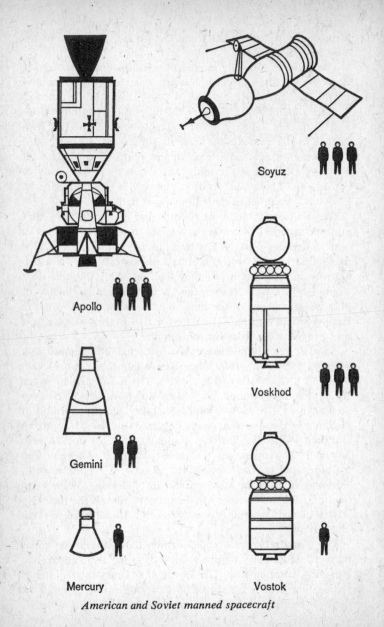

Soyuz

Apollo

Voskhod

Gemini

Mercury

Vostok

American and Soviet manned spacecraft

they are awaiting a larger booster, even more powerful than the American Saturn V, or whether it is their intention to assemble their moon expedition in earth orbit, remains to be seen.

In September 1968 the unmanned probe Zond 5 carried animal life including turtles round the moon and back to a splashdown in the Indian Ocean, the first time the Soviets had used a sea recovery. The turtles were none the worse for their trip. In November a second probe, Zond 6, repeated the mission, returning this time to a land recovery. Though a Russian attempt to land a man on the moon before Apollo 11 seemed unlikely, since there was no evidence to suggest that they had yet tested a moon landing craft, it could not be completely ruled out.

The Apollo 9 mission launched into earth orbit on 3 March 1969 after a three-day delay caused by astronaut colds was the first manned flight for the lunar module or the 'tissue-paper spacecraft', as the mission commander James McDivitt described it. The purpose of the mission, he said, was 'to see if this whole gigantic mess would be able to fly to the moon and land there'.

On 5 March, James McDivitt and 'Rusty' Schweickart clambered through the tunnel linking the command module nicknamed 'Gumdrop' with the lunar module 'Spider' while David Scott remained in the command module. Six hours later they started up the main lunar module descent engine and the linked Spider–Gumdrop combination was pushed into a higher orbit but a medical problem had developed which threatened to become serious. Climbing through the tunnel, Schweickart had suffered an attack of nausea. (A space-walk planned for the next day was at first cancelled but later Schweickart managed a thirty-seven-minute trip outside the LM.) After further tests of the LM descent engine the decision was made to take the LM on its first solo flight.

On 7 March, McDivitt and Schweickart separated Spider from Gumdrop and for six hours the two spacecraft flew separately, as far as 100 miles apart. Re-docking the two craft presented no problems and the lunar module was now ready for lunar flight. The Apollo 10 mission, scheduled for 18 May, would take the complete Apollo spaceship to the moon for the final test before the landing.

The crew for that mission was Thomas Stafford, commander; Eugene Cernan, lunar module pilot; and John Young, command module pilot. Launch was exactly on time and within three hours they had left earth orbit and were starting man's second journey to the moon. They carried with them a colour television camera and from the start it was clear that they were not only enjoying the expedition but were determined to share their experiences with their anxious audience around the world. The laconic humour with which they reported their progress dispelled once and for all any doubts there may have been about whether astronauts were really human.

As they journeyed out over quarter of a million miles of space towards the moon they sent back live colour television pictures of the cloud-strewn planet earth. The outlines of the continents could be seen clearly. India and South America appeared ochre while the oceans were a deep ultramarine against the blackness of outer space. For the first time millions saw, through the eyes of a space traveller, the 'Good Earth', as Frank Borman had described it.

On their way to the moon the Apollo 10 crew detached the command and service modules from the top of the last stage of the Saturn rocket, turned the CSM around and docked it with the lunar module. For the first time the 'transposition and docking' manoeuvre was carried out with the complete Apollo spacecraft. They then restarted the Saturn engine and, its task now complete, blasted it out into orbit around the sun. They reached the moon on 21 May and, while they were behind it, the service module engine was fired to put them into lunar orbit. The next day the crucial test of the lunar module was begun.

At 4.40 p.m. (BST) Stafford climbed through the tunnel into the LM, codenamed 'Snoopy' for this mission. At 5.25 p.m. Cernan joined him, leaving John Young alone in the command module, 'Charlie Brown'. As they were preparing to undock the two craft, a hitch developed. Not all the air in the connecting tunnel had been evacuated as it should have been and to keep the two craft linked at this stage a vacuum seal was necessary.

This seal was not now perfect and the two spacecraft were slightly twisted out of the correct alignment. As they disappeared behind the moon, ground control told them not to undock if the problem became worse. The world waited tensely for them to reappear at the beginning of the next orbit.

'We're down among 'em,' came Stafford's voice across 239,000 miles of space. Safely undocked, Snoopy was beginning a 3,500-mile-an-hour swoop to within 30,000 feet of the landing site chosen for the first American on the moon. Taking both cine and still photographs as fast as they could, they tried also to give a running commentary on the nearest approach yet to the hostile lunar terrain. Unfortunately a communications failure reduced their voices to a mere whisper, but as they flew directly over the July landing site Cernan's voice could be heard: 'It's smooth, like wet clay,' he said.

Another momentary hitch occurred a few hours later when Stafford and Cernan jettisoned the lower half of the lunar module, simulating the launch from the lunar surface. A switch inadvertently left in the wrong position caused an unexpected change in the attitude of the spacecraft. 'Son of a bitch,' said an angry Cernan but the difficulty was soon passed. The top half of Snoopy carried the two astronauts back up to John Young in Charlie Brown where the tired crew were safely reunited.

After a well-earned rest they prepared the spacecraft for the home journey. Both halves of Snoopy had now been discarded and the next day they fired the service module engine to start the long trip back.

Their splashdown in the early hours of 26 May was within two miles of the target area and in sight of the recovery aircraft carrier. Taking many of the risks in return for little of the glory, Stafford, Cernan and Young had finally prepared the way for the first man on the moon and the fulfilment of President Kennedy's promise. Even as they journeyed out a giant crawler was edging its massive load towards launch pad 39 A at Cape Kennedy. The Apollo 11 moonship was on its way to the launch pad.

Cape Kennedy
16 July 1969

Equipment

Twenty years ago the dusty country road that branched off the busy highway linking the cities of America's north-eastern seaboard with the Florida sun led across two stretches of water to a marshy bulge on the otherwise straight peninsular coastline. Its name was Cape Canaveral and behind it, between the two channels of the Indian and Banana Rivers, lay the low flat sprawl of Merritt Island, home of pelicans, egrets and herons. Today, Merritt Island is the nesting ground of giant space birds, the Saturn V moon rockets. Hatched at the rate of one every two or three months, they roar off into the Atlantic sky and in a few minutes are gone for ever.

It took eight years and a billion dollars to turn the wilderness of Merritt Island into the John F. Kennedy Space Centre (KSC) for the Apollo programme. In the middle of this moonport is Launch Complex 39, dominated by the largest enclosed space in the world, the vehicle assembly building (VAB). It is said that clouds gather in the top of the VAB if the doors are left open at night. Standing 525 feet tall on an area the size of four football pitches, this 109-million-dollar hangar can house the separate components that go to make up four Saturn Vs as well as the completed vehicles. The foundations were laid in July 1963 and by February 1966 the VAB was ready and equipped for the moon rockets. The Saturns are assembled inside the VAB on massive mobile launching platforms costing another $101 million.

The first stage of the Apollo 11's Saturn V, the S-IC built by Boeing in New Orleans, was thirty-three feet in diameter and 138 feet long. Delivered by sea-going barge to the Kennedy

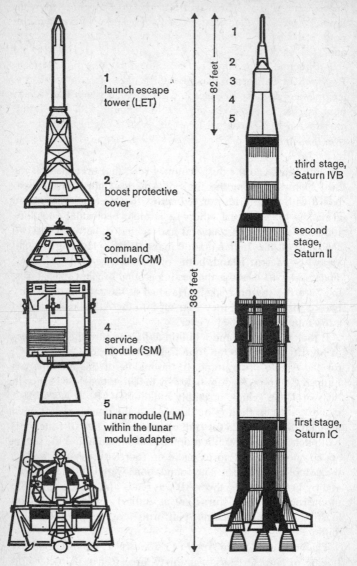

1
launch escape
tower (LET)

2
boost protective
cover

3
command
module (CM)

4
service
module (SM)

5
lunar module (LM)
within the lunar
module adapter

82 feet

363 feet

third stage,
Saturn IVB

second
stage,
Saturn II

first stage,
Saturn IC

The Apollo 11 Saturn V and mooncraft

Space Centre, it weighed 288,750 pounds. Fully fuelled with liquid oxygen and kerosene, this increased to over five million pounds, its five FI engines developing a thrust of over seven and a half million pounds.

The second stage, the S-II assembled by North American Rockwell in California, was also thirty-three feet in width. Its length was just over 81½ feet and its unladen weight of 79,918 pounds rose to just over one million pounds with the liquid oxygen and hydrogen fuel load. Its five J 2 engines developed over one million pounds of thrust. The S-II was also delivered by barge.

The third stage, the S-IVB built by McDonnell Douglas in California, just under twenty-two feet in diameter and a little over fifty-eight feet in length, weighed 25,000 pounds dry and 260,523 pounds fuelled with liquid oxygen and hydrogen. A single restartable J 2 engine provided a maximum thrust of 203,779 pounds. The S-IVB was delivered by an aircraft called the Super Guppy which looks like a balloon with two short wings. A smaller version is appropriately called the Pregnant Guppy. The engines of all three stages were test-fired before delivery; the S-IC and the S-II stages at the NASA Mississippi Test Facility, the S-IVB at McDonnell Douglas.

The erection of the Apollo 11 Saturn V began when a diesel-powered transporter was driven to the mobile launcher parking area. The transporter matched the Saturn V in mammoth proportions. In appearance it was like half a football pitch borne twenty feet aloft by four ten-foot-high double tank tracks, one at each corner. Each link of track was seven and a half feet long and weighed one ton. Unladen, it weighed six million pounds and possessed a majestic maximum pace of one mile an hour. NASA have spent over $12 million on these giant crawlers. Moving beneath one of the launchers, the crawler's sixteen twenty-inch-diameter mobile hydraulic jacks lifted the 160-foot-long, 135-foot-wide launcher platform clear of its pedestals. The 445-foot-high structure was then taken through one of the doors of the VAB to await the arrival of the first stage of the Saturn. NASA have built three mobile launchers for the Apollo project, each weighing twelve million pounds unladen, and they serve both as mobile launch platforms and as fuelling, power and

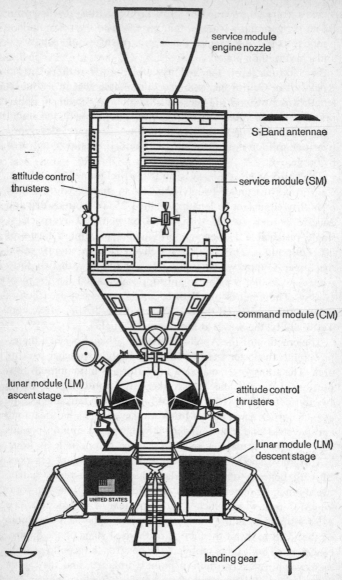

service module
engine nozzle

S-Band antennae

attitude control
thrusters

service module (SM)

command module (CM)

lunar module (LM)
ascent stage

attitude control
thrusters

lunar module (LM)
descent stage

UNITED STATES

landing gear

The complete Apollo 11 spacecraft

communications towers for the Saturn while it is on the ground. The platform area is about half an acre and the tower supports nine retractable service arms which carry the umbilical lines. As the moment of lift-off approaches, the arms are retracted. At the 320-foot level, serviced by a sixty-mile-an-hour lift, the Apollo access arm provides access to the command module for the astronauts and the 'close-out' team who help them aboard. This arm is retracted to about three feet from the spacecraft forty-three minutes before lift-off and can be rapidly reconnected up to five minutes before launch in the event of an emergency. (After T minus 5 minutes the launch escape rocket system would have to be used.)

Transported in February to the Kennedy Space Centre by the NASA barge *Poseidon*, the S-IC stage of the Apollo 11 Saturn was brought to within a few yards of the VAB along a canal. From the canal basin the S-IC was taken into the VAB for two days of intensive checks before being hoisted into its vertical position on the mobile launcher by a 250-ton capacity bridge crane. Within a week, all the umbilicals were attached, including lines connecting every single part of the S-IC to the Space Centre's computer complex. The next few weeks were taken up with extensive electrical and mechanical checks. Meanwhile the other two stages of the Saturn V were undergoing check-out in another part of the building. These tests completed, the S-II stage was lifted into its place atop the S-IC and the two stages mechanically mated. On 5 March the third stage, the S-IVB, was added. Already attached to the top of the S-IVB was the instrument unit. This was shaped like a table-napkin ring, was built by IBM, and was the Saturn's onboard computer. Its job was to control and navigate the first three stages of the Saturn from lift-off to earth parking orbit. A day after all the umbilicals linking the three stages of the Saturn with the launch tower were connected, the full electrical power was turned on and three weeks of check-out of the now complete Saturn V began. At the end of this period the launch vehicle was ready to accept an Apollo spacecraft. This came in three parts: the command module (CM), the service module (SM) and the lunar module (LM).

The command module is a cone 11 feet 5 inches high with

a base diameter of 12 feet 10 inches. It weighs 12,250 pounds at lift-off with the three astronauts aboard and is their home for the greater part of the journey to the moon and back. It is the only part of the moonship which returns to earth. It is divided into three compartments: forward, crew and aft. The forward compartment is the relatively small area at the apex of the cone. Below this is the crew compartment which occupies most of the centre section of the CM. The aft compartment takes the rest of the space around the edge of the base of the module.

During launch and re-entry the CM is orientated so that the base is pointing downwards, and at these stages of the mission the three members of the crew lie on their backs in their couches with the side access hatch behind them. The astronauts spend much of the journey on these couches but they can leave them and float around the cabin once they are weightless. Folding back the seat section of the centre couch makes room for two astronauts to 'stand' in the middle of the capsule at the same time. For rest periods there are two sleeping-bags attached beneath the left and right couches. Sleeping astronauts are tethered to these bags so that they do not float around in their slumbers.

Food, water and other equipment are housed in bays which line the spacecraft walls. The cabin is normally pressurized at five pounds per square inch, which is about one-third of atmospheric pressure at sea-level. At launch they breathe an air mixture but this is gradually replaced by pure oxygen. The cabin temperature is kept at 75° Fahrenheit. This environment enables the crew to spend much of the time out of their space-suits, though these are worn during critical phases such as launch, re-entry, docking, and crew transfer to the lunar module.

The skin of the CM consists of two layers: the inner pressure shell and the outer heat shield. The inner shell is made up of an aluminium sandwich with a honeycomb structure between two layers of sheet metal. The honeycomb structure makes for rigidity and lightness and its thickness varies from one and a half inches at the base to a quarter of an inch near the apex. The outer shield is also made up of a honeycomb sandwich but in this case a steel alloy is used. Its thickness varies from a half to two and a half inches. Between the inner and outer layers, fibrous insulation provides additional heat protection both from the temperature extremes of outer space and the heat of re-entry.

The heat generated by air friction during launching is absorbed by the boost protective cover which is jettisoned after just over three minutes of flight, by which time the Saturn has cleared the denser layers of the atmosphere. The temperature of the protective cover can reach about 1,200° Fahrenheit, and for this reason it is made of the best insulation materials available, fibreglass and cork, and fits over the CM like a glove.

The forward compartment around the tunnel which gives access to the LM is separated from the crew compartment by a bulkhead and covered by the forward heat shield. It is divided into four segments which contain the landing parachutes as well as two reaction control engines, part of the CM's attitude control system which orientates the craft in space.

The aft compartment consists of twenty-four bays ringed around the base of the CM. Included in these bays are ten reaction-control thruster engines, fuel for these engines, water tanks and a number of instruments. The CM-SM umbilical (the plumbing connecting the two modules) is also attached to the aft compartment.

The crew compartment has a habitable volume of 210 cubic feet. The environment control subsystem (ECS) maintains temperature and pressure. The compartment also contains the controls and displays for flying the spacecraft, and the crew couches. It has two hatches and five windows. A number of bays or cupboards provide space for equipment.

The side hatch is used for getting in and out of the CM, while the forward hatch is used for transfer into and out of the lunar module. The side hatch is twenty-nine inches high and thirty-four inches wide, with a nine-inch-diameter window in the middle. A single handle on the inside opens the twelve latches which secure the hatch, a modification which followed the fire in January 1967. The latches are so designed that pressure on the hatch serves to increase the locking pressure of the latches. Should the single handle fail, the latches can be opened and closed by hand. The hatch can also be opened from the outside by an emergency wrench carried on board as well as by ground personnel. The hatch handle also operates the hatch in the boost protective cover, enabling the crew to leave the spacecraft before launch, for example during training sessions and in emergency until five minutes before lift-off. On the launch pad the ground crew can easily close the side hatch by simply pushing it. Up in space the crew can close the hatch from the inside by pulling a handle near the lower hinge which swings the hatch inwards. As a further safeguard, the cabin's tool kit contains a set of three jackscrews which in emergency would keep the hatch secured in the event of failure of the latches. A valve built into the side hatch allows the crew to vent the cabin atmosphere. This takes about one minute and is carried out before the hatch can be opened in the vacuum of outer space, though this should not be necessary during a moon mission. This valve can also be operated from the outside.

The detachable forward hatch is circular, thirty inches in diameter, and gives access to the tunnel connecting the CM with the LM. A single handle similar to that on the side hatch operates six latches. A valve enables the pressure in the CM and LM to be equalized before this hatch is opened. There are also provisions for opening the hatch by hand in emergencies.

The CM has five windows. One, as already described, is in the centre of the side hatch. On either side of this hatch are two thirteen-inch-square windows which are also next to the left- and right-hand couches. They are used for observation and photography. It is through these windows that television pictures of the moon and earth are taken. Above these two windows and slightly nearer the side hatch are two forward-looking windows used during rendezvous with the LM. These are triangular in shape, eight by thirteen inches in size. They are used by the astronaut seated on the right, the command module pilot (Collins on the Apollo 11 mission), and on the left, the mission commander (Armstrong). The astronaut seated in the centre couch, the lunar module pilot (Aldrin), uses the window in the side hatch. (On the Apollo 11 mission these seating positions were changed before the transposition and docking manoeuvre, when the order from left to right became: Collins, Armstrong, Aldrin.)

The windows are triple-paned: the two quarter-inch-thick inner panes are one-tenth of an inch apart, while the outer pane is nearly three-quarters of an inch thick. Each pane is coated on both sides to reduce reflection and filter out infra-red and ultra-violet radiation. The outer panes can withstand temperatures up to 2,800° Fahrenheit, well above the temperature of re-entry in the region of the spacecraft windows. Aluminium shades can be fitted by the crew to cut out light.

The interior of the CM is lined with storage bays containing enough equipment for fourteen days in space as well as much of the electronics and other apparatus needed for the operation of the spacecraft. The lower bay at the foot of the centre couch contains the guidance and navigation electronics, a sextant and a telescope, the computer (in fact two computers, the second checks the first) and the computer keyboard. It also houses most of the telecommunication equipment used to link the CM with the manned spaceflight network (MSFN) on the ground and with the LM when it is separated. Batteries, food and other equipment are stowed in this bay. The left-hand equipment bay beside the left-hand couch contains an important part of the environment control subsystem (ECS) including the environment control unit (ECU). There is space in this bay to stow the forward hatch when the tunnel into the LM is in use. The left-hand forward bay, in front of the left-hand couch, houses other ECS equipment including the water delivery unit. This bay also serves as a clothing store. The right-hand equipment bay beside the right-hand couch contains food supplies as well as waste management systems. The right-hand forward bay is used to store medical kits, survival gear and the camera equipment carried on a mission. In the aft equipment bay, beneath the couches, are elements of the suits, helmets and life support systems for use on the surface of the moon and for critical moments in flight. The docking probe is also stowed in this bay when the CM and LM are linked.

For most of the mission the command and service modules

are linked to form one unit, generally referred to as the command and service module (CSM). They only separate minutes before the CM enters the atmosphere prior to splashdown. The connexion is made by three stainless steel straps which are severed by small explosive charges at separation. Also linking the CM with the SM is the umbilical which carries power, water and oxygen between the two modules. At separation the electrical circuits are switched off (deadfaced) and valves close the tubing before a guillotine mechanism cuts the umbilical.

The command module carries only 270 pounds of fuel, which is used to operate attitude-changing reaction control thrusters. The main fuel load, about 42,000 pounds, is kept in the propellant tanks of the service module. Once the last stage of the Saturn has been jettisoned, it is the SM's function to make speed and course changes on the way to the moon and back. The SM is 24 feet 7 inches in length and 12 feet 10 inches in diameter. With its full fuel load it weighs over 51,000 pounds, with the propellant expended only 11,500 pounds. The main engine which is fired to slow the spacecraft down when it reaches the moon and so put it into lunar orbit, and to push the craft on its journey from lunar orbit back to earth, has a restartable single rocket engine developing 20,500 pounds of thrust. The SM also contains the greater part of the oxygen and water supply for the mission, as well as much of the electrical power, the ECS, and a small part of the communications system. The SM also has four sets of reaction control thrusters at ninety-degree intervals around its diameter.

The interior of the SM is divided into six sectors by panels radiating from a central tube which houses the main engine. Four of these contain fuel tanks for the SM engine, while a fifth contains oxygen tanks for the CM and for the fuel cells which provide electricity and drinking water. The sixth sector is basically empty to allow additional equipment to be carried on later missions. Also distributed among the five sectors in use are elements of the ECS, three fuel cells and part of the spacecraft communications system, as already mentioned.

The command and service modules are constructed by North American Rockwell in California by engineers dressed like surgeons, working in gigantic 'clean rooms' where conditions rival those in hospital operating theatres. The modules are delivered to Cape Kennedy by air.

The lunar module, the sole purpose of which is to land two men on the moon and bring them safely back to the CSM, is built by the Grumman Aircraft Corporation in New York State. The LM is 22 feet 11 inches high and has a maximum diameter with the legs extended of 31 feet. With its full fuel load and the two crewmen aboard, it weighs 33,205 pounds, of which nearly 24,000 pounds is accounted for by the propellants.

The LM is made up of two stages: the descent stage is essentially a platform with four legs and the rocket engine used to land the module on the moon, while the ascent stage consists of the crew cabin and the rocket engine which, using the descent stage as its launch pad, carries the crew back up into lunar orbit to the waiting CSM. The LM is designed and equipped to operate for only forty-eight hours as a separate craft, allowing a maximum stay-time on the moon of about thirty-five hours.

The ascent stage is 12 feet 4 inches high, 14 feet 1 inch wide, and weighs 4,804 pounds before fuelling, which brings it up to a launch pad weight of just over 10,000 pounds. It is similar to the command module in that it consists of the crew compartment of the LM with its own environment control systems, attitude control thrusters, communications, navigation equipment and storage bays. Controls and displays for the main engine and attitude thrusters in the two-man cabin are duplicated to allow either member of the crew to fly the craft.

The descent stage is a modified octagon in shape standing 10 feet 7 inches high with the landing legs extended; fully fuelled, it weighs just over 22,500 pounds. The octagon measures 14 feet 1 inch corner to corner. It is similar to the service module in that it is mainly made up of the descent engine with its fuel tanks. The descent stage also contains water and oxygen supplies as well as the scientific experimental packages which the astronauts leave behind on the lunar surface. A more detailed description of the lunar module follows in subsequent chapters. Since the LM is not aerodynamically streamlined, it is housed in a four-panelled spacecraft-lunar module adapter during the launch phase of the mission and is situated between the third stage of the Saturn and the CSM. Once on the way to the moon, the

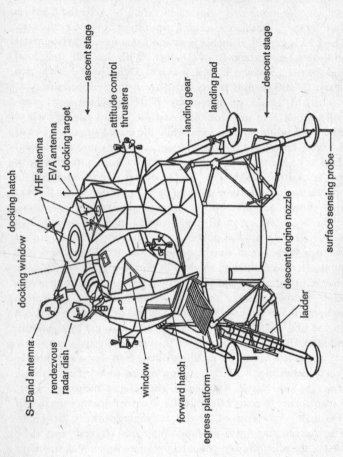

ascent stage

descent stage

attitude control
thrusters

landing gear

landing pad

docking target

EVA antenna

VHF antenna

docking hatch

docking window

S—Band antenna

rendezvous
radar dish

surface sensing probe

descent engine nozzle

ladder

window

forward hatch

egress platform

The Apollo 11 lunar module (L M)

correct configuration is achieved by the transposition and docking manoeuvre. It is also for safety reasons that the CSM and LM are not launched in the configuration in which they will travel to the moon: the CM with the launch escape tower (LET) has to be at the top before and during lift-off so that the astronauts can blast themselves clear of the Saturn in any emergency.

The thirty-three-foot rocket and tower assembly comes into use automatically if triggered by the emergency detection system, or can be started by the astronauts any time from T minus 5 minutes until just over three minutes into the mission after launch. Three seconds of 147,000-pound thrust from the launch escape rocket carries the command module clear of the rest of the spacecraft to a parachute landing after a very bumpy ride. Not surprisingly, the system has only been tested with unmanned modules.

Before any of the Apollo modules are launched they have 'flown' the complete mission a number of times during extensive ground tests. The final tests are carried out at Cape Kennedy. As soon as they arrive the modules are taken into the manned spacecraft operations building (MSOB) situated some three miles south of the VAB. The Apollo 11 command and service modules arrived as one unit, the lunar module separately. Again under operating-theatre conditions, each module was examined minutely before the CSM was mated to the LM, now encased in its adapter housing. In the altitude chamber of the MSOB tests were carried out in conditions simulating an altitude of 200,000 feet with the astronaut crew aboard. The complete Apollo 11 spacecraft was then transferred to the VAB where its was hoisted on top of the Saturn V and mated to the S-IVB stage. Umbilicals were then attached linking the spacecraft with the mobile launcher tower. The attachment of the launch escape tower to the top of the command module completed the moonship.

Through the umbilical arms every single part of an Apollo craft is linked to the acceptance check-out equipment (ACE) computer housed in the MSOB. During the preparations for the launch of a Mercury spacecraft, eighty-eight separate systems were tested individually by an engineer on the spot, each of these checks taking several minutes. The ACE computer

53

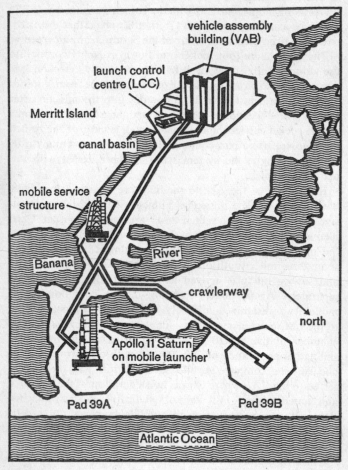

Complex 39 at Cape Kennedy, 16 July 1969

checks out several thousand Apollo spacecraft systems, involving over one and a half million working parts, every second.

On 21 May the Apollo 11 Saturn was ready for the first part of its journey to the moon: just over three miles along the 131-foot-wide crawlerway roadbed to Launch Pad 39A at a cosmic speed of less than one mile an hour. This short piece of highway cost NASA over $24 million, or $1,500 a foot. As the crawler carried its precious load with its two 2,750-horsepower diesel engines, the roadbed sank several inches under its tracks. Two further 1,065-horsepower engines powered the jacking, steering, lighting and electronic systems on board, including a highly sensitive levelling device which kept the Saturn vertical to within ten seconds of arc. A tilt of as little as two inches at the top of the launch escape system was automatically corrected by this device, which also kept the rocket vertical as it was carried up a gentle ramp on to the launching pad.

Launch Pad 39A is octagonal in shape and a quarter of a mile square. In the centre of the pad is a reinforced concrete flame trench fifty-eight feet wide and forty-two feet deep. At one end of this trench a flame deflector mounted on rails directs the fiery blast of the departing Saturn. Depositing its load directly above this trench, the crawler then transported a 402-foot-tall mobile service structure.

The mobile service structures are towers of stressed steel which provide all-round access by means of platforms on five levels enclosing the Saturn on the mobile launcher. They have two high-speed lifts. The tower is normally kept in position until about ten hours before lift-off (T minus 10 hours). The towers weigh nearly nine million pounds and cost NASA over $16 million to build.

Once it has reached the pad, a Saturn is reconnected to a second computer complex centred in the launch control centre (LCC). This constantly monitors the proper functioning of the five million working parts of the Saturn systems in the same way as the ACE computer was used to check out the Apollo spacecraft. It is also linked with the onboard instrument unit which carries out its own pre-launch check-out of the Saturn. The LCC is adjacent to the VAB and contains four identical firing rooms enabling work on four Saturns to proceed simultaneously. Each firing room has 450 consoles giving access to the computer, as well as fifteen giant display systems and monitors attached to sixty television cameras positioned around the Saturn. Though aided by the highly automated computer systems, it still takes a team of 550 men (backed up by a further 5,000 technicians) to launch a Saturn V.

Men

In 1966, long before the first nut of either the Saturn V or the Apollo spacecraft had met the first bolt, the crew of the Apollo 11 mission were making their first trips into orbit. Neil Armstrong, Apollo 11 commander, captained the hazardous Gemini 8 mission in March. Completing the first ever space docking manoeuvre, he ran into trouble when the two craft got out of control and the trip ended with an emergency splashdown. In July, Michael Collins completed eighty-eight minutes of space-walking on the Gemini 10 mission, which also included two link-up manoeuvres. In November, Edwin Aldrin also experienced both space-walking and docking on the Gemini 12 trip. Apart from 1966, the three Apollo 11 astronauts share another common date: they were all born in 1930.

Neil A. Armstrong was born on 5 August 1930. He spent part of his childhood building model aeroplanes at his home in Wapakoneta, Ohio. At the age of five he persuaded his father, an auditor working for the state government, to take him for a ride in a visiting ancient Ford Trimotor plane. (He started learning to fly at the age of fourteen and got his wings two years later.) As a Navy pilot between 1949 and 1952 he was involved in the Korean war and flew seventy-eight combat missions. On one sortie he was shot down but parachuted to safety on friendly territory. Returning to civilian life, he took a course in aeronautical engineering at Purdue University from which he graduated in 1955. He immediately joined the Lewis Research Centre where, as a test pilot, he flew ever faster and higher. In April 1962 he reached a height of forty miles in an X-15 rocket plane and has flown at nearly 4,000 miles an hour, five times the speed of sound. He has logged over 4,000 hours of flying time, or nearly half a year. That same year he was selected by NASA to become a trainee astronaut. Following his Gemini flight, Armstrong was involved in development work on the lunar landing training vehicles. It was in 1968, while piloting one of these flying bedsteads, that he had yet another narrow escape when he parachuted clear of the crashing machine. Armstrong, probably the world's most experienced pilot, is used to getting himself out of tight corners.

Edwin E. 'Buzz' Aldrin, lunar module pilot, was born in Montclair, New Jersey, on 20 January 1930, the son of a US Air Force colonel. His father sent him to West Point, the US military school, from which he graduated in 1951 with a degree in science, placed third in his class of 475. He then joined the Air Force, flew sixty-six combat missions in Korea, and was decorated. In 1963 he completed work for a second degree in astronautics at Massachusetts Institute of Technology, and soon found himself at NASA's Manned Spaceflight Centre in Houston, Texas, where his job was to coordinate Air Force participation in the Gemini experiments. In October that year he was one of a third team of fourteen new astronauts selected by NASA. After flying on the last of the Gemini missions, Aldrin trained intensively for the Apollo series. Like Armstrong, he too has spent much time learning how to fly the lunar module. Colonel Aldrin has logged 3,500 hours flying time in jets and helicopters.

Michael Collins, command module pilot, was born in Rome on 31 October 1930. His father, of Irish descent, was then military attaché at the US embassy. Collins had a military education, graduating from West Point with a science degree in 1952. He then joined the USAF, in which he holds the rank of lieutenant-colonel. For a number of years he acted as an experimental test pilot at the USAF's Edwards Air Force Base in California, before being picked by NASA as a trainee astronaut along with Aldrin in 1963. Following his flight on the Gemini 10 mission, he was selected as command module pilot for Apollo 8 but spine trouble confined him to hospital and his place was taken by James Lovell. Collins has 4,000 hours of jet flying experience behind him.

Viewed as 'space hardware', the astronauts' other statistics are as follows: Armstrong: blond hair; blue eyes, height 5 feet 11 inches; weight 165 pounds; salary $30,054. Aldrin: blond hair; blue eyes; height 5 feet 10 inches; weight 165 pounds; salary $18,623. Collins: brown hair; brown eyes; height 5 feet 11 inches; weight 165 pounds; salary $17,147. Their government salaries are more than doubled by publishing rights which all the astronauts share amongst themselves.

For every hour of the Apollo mission Armstrong, Aldrin and Collins spent over five in training, a total of more than 2,000 hours. In addition, each crew member spent many more hours taking physical exercise, studying and attending briefing and review meetings. It takes twelve months to train an Apollo crew, the members of which must normally have at least eighteen months of astronaut experience behind them. Their training ranged from hundreds of practice sessions for every stage of the flight and various possible emergency procedures to geological field trips in Iceland and jungle survival courses in Panama. The Apollo 11 crew were not only expected to fly to the moon, land there and return safely, but also to take on the roles of scientists, explorers, navigators, photographers, television performers and, if the need arose, mechanics.

In charge of preparing the Apollo 11 crew for flight was Donald Slayton, one of the first team of seven astronauts chosen by NASA. He was grounded for medical reasons but subsequently became director of astronaut training. The training programme was divided into four main areas: spacecraft testing, briefings and reviews, 'flying' in spacecraft simulators, and specific training for other phases of the mission. The Apollo 11 back-up crew, James Lovell and William Anders, both veterans of Apollo 8, and a newcomer, Fred Haise, joined in on all but the last few days of training sessions. A third crew substituted when the tests were purely for the benefit of technicians checking the spacecraft systems.

The Apollo 11 crew spent more than 160 hours in the command and lunar modules at the factories where they were built and at the Kennedy Space Centre where they were prepared for flight. They also attended more than 135 hours of briefing on the modules from the manufacturers and NASA scientists. At Cape Kennedy they also practised docking the two spacecraft, getting out of both modules and escaping from the launch pad in an emergency, and took part in the flight readiness test and a full-scale demonstration countdown which ended on 3 July.

The crew attended a total of seventeen separate briefings, reviews and meetings covering every step of the Apollo 11 flight plan. Before their flight each member of the crew spent more than 400 hours 'flying' command and module simulators at the Cape Kennedy Space Centre and the Manned Space-flight Centre in Houston, Texas. Linked to computers, these flight simulators allow astronauts to practise landing on the moon, docking and other vital manoeuvres in highly realistic conditions.

In addition to simulator sessions, Armstrong flew the lunar landing training vehicle (LLTV). This was the successor to the lunar landing research vehicle (LLRV) which gave Armstrong some nasty moments in May 1968 when he had to use his ejection seat at low altitude to escape a crash as the craft went out of control. A month before launch day, between 14 and 16 June, Armstrong flew the wingless experimental LLTV eight times for a total flight time of forty minutes. Three of these trainers have been built by Bell Aerosystems at a cost of $2·5 million each. A 4,200-pound thrust jet engine lifts five-sixths of the craft's 4,051-pound weight. The remaining one-sixth of the vehicle's weight (what it would weigh on the moon) is lifted by two 500-pound thrust rockets. Attitude control is governed by sixteen small rocket engines clustered in groups of four as they are on the LM. This enables the craft to pitch, roll and yaw. Aldrin also made some flights, using a similar but tethered craft.

The need for astronauts to learn to walk on the moon has led to the construction of training equipment no less bizarre than the LLTV. In one such rig a man is suspended from one arm of what looks like a giant mobile conceived by some futuristic sculptor. At one-sixth of his earthly weight he then tries to progress across various patches of rocky ground pre-pared to look like the lunar surface. Experiments like this revealed that, while it is easier to lift a foot under lunar gravity conditions, putting it down again can present problems. In another training rig the man is lifted horizontally in a harness so that he swings like a pendulum. He then tries his skill at walking along the face of a vertical wall.

The Apollo 11 astronauts also practised getting in and out of their spacesuits and command and lunar module mock-ups aboard aircraft flying in a parabolic loop. This provided about a minute of zero-G (weightlessness) experience in the cabin. Weightlessness can also be simulated under water and the crew practised crawling through the CSM-LM tunnel and spacewalking wearing their pressure suits in immersion tanks fitted out with mock-ups of the spacecraft.

In April, Armstrong and Aldrin had carried out a 2½-hour full dress rehearsal of their moon walk in full kit at the Manned Spaceflight Centre in Houston. Going through all the motions of getting out of and back into the lunar module, sample collection, and setting up the experiments to be left on the moon, they were only spared one-third of their equipment load as they toiled away under earthly gravity. On the moon itself their 180-pound load would only weigh thirty pounds under the one-sixth lunar gravity. On 24 May all three members of the Apollo 11 crew practised getting out of the command module following splashdown. In the Gulf of Mexico they went through the strict quarantine routine to be followed between their splashdown and transfer by air to the Lunar Receiving Laboratory at Houston – an 8·5-million-dollar precaution against the remote possibility of lunar bugs.

Meanwhile elaborate quarantine precautions of a different sort were coming into force. The crew were being kept from contact with all but the most essential people to lessen the risk of some earthly complaint ruining the mission as nearly happened on the four previous manned Apollo flights. At an early press conference the astronauts sat behind screens with the air conditioning blowing their germs towards the press in preference to vice versa. At a later conference only a television team, the members of which had passed a medical test, was allowed in the same room as the Apollo crew. Dr Charles Berry, the astronauts' chief doctor and medical director of NASA, even scuttled a presidential request to dine with the crew on the eve of the launch. The big handshake would have to wait until splashdown, though at this stage is was not quite clear how this would be managed and one report suggested that Dr Berry was threatening to put President Nixon in quarantine along with the crew. While the Apollo crew were obviously disappointed by Dr Berry's decision, they kept their feelings to themselves. Not so Frank Borman, commander of the Apollo 8 mission, who had just returned from a goodwill visit to the Soviet Union; at a press conference at Cape Kennedy on 13 July he strongly criticized the medical decision cancelling the dinner.

Another arrival from Russia that same day was a news flash: Luna 15, an unmanned probe, was on its way to the moon. Few details were given by the Soviet news agency, Tass, whose enigmatic dispatch merely stated that the purpose of Luna 15 was 'to explore space in the vicinity of the moon'. Rumours that the Russians were about to attempt the automatic retrieval of a sample of moon rock had been circulating in Moscow for some weeks. It was even suggested that two failures in the attempt had already occurred when the launch vehicles had exploded on their pads. Luna 15, it seemed, was due to reach the moon the day after the Apollo 11 launch. Not until then was it known whether Luna 15 was going to try to upstage the American landing.

Following the successful completion of the demonstration countdown, final tests of the Saturn V, now enshrouded on Launch Pad 39A by the giant mobile service tower, went smoothly ahead. Only a minor leak, detected on 10 July, in the helium gas system used to pressurize the liquid oxygen in the first stage gave engineers anything to worry about, but the fault was quickly cleared up. The next day Dr Berry gave the astronauts their final medical test and declared: 'Their medical state looks good. The Apollo 11 crew is ready for flight.'

After a 'relaxing' week-end flying and swimming, the crew were now also resident at Cape Kennedy in the manned space-flight operations building, seven miles south of the launch pad. They spent Monday 14 July going over the critical moon landing and take-off for the last time in simulators. Tuesday 15 July, the eve of the launch, was a day of complete rest, an end to the six-day weeks and twelve-, sometimes fifteen-hour days of rigorous training. The pace of the Apollo programme may well have been the reason why some crews on earlier missions showed signs of fatigue when it came to the flight itself, and Dr Berry in particular was anxious that the Apollo 11 crew should be fresh from the start.

In the evening the crew dined in the MSOB. The menu: sirloin steak, potatoes, buttered asparagus, cottage cheese, fruit and 'a beverage', and to share it with them just seven guests: the two stand-by crews and the man who trained them, Donald Slayton. As the sun set on Cape Kennedy, enormous arc lights lit up the giant bird perched on the launch pad as the digital countdown clocks around the space centre ticked off the seconds. By now the mobile service tower had been moved along the crawlerway to a park a mile or so away. The countdown, the most trouble-free to date, was ahead of schedule. Shortly after midnight at the Cape, the countdown now at T minus 9 hours, the Saturn V was readied for its massive fuel load. First of all, the tanks of the three stages were cooled with liquid nitrogen so that the liquid oxygen and hydrogen would not boil off as it was fed into them at more than 250° below zero. Already the kerosene was aboard the first stage, but now from storage tanks around the perimeter of the pad the fuel began pouring in through 15,000 feet of ten-inch piping at a rate of 10,000 gallons a minute. It took nearly five hours to fill the Saturn's tanks and while this was going on nobody was allowed within half a mile of the rocket.

Blast-off

In the pre-dawn hours of Wednesday 16 July 1969, as the astronauts slept, the seconds of the countdown slipped quietly away into the warm night at Cape Kennedy. So far, not a hitch. To technicians, long used to fretful holds and agonizing postponements, it was almost a miracle. 'When we wheel out one of the rockets to the launch pad,' remarked von Braun, 'I find myself thinking of all those thousands of parts, all built by the lowest bidder, and I pray that everyone has done his homework.' Just before the sun came up over the Atlantic, the crew had the traditional pre-launch breakfast of steak, egg, toast, coffee and orange juice. Then, kitted up in their spacesuits, they climbed aboard the transfer van for the seven-mile journey from the MSOB to Launch Pad 39A. After a thirty-two-second ride in the high-speed lift on the mobile launcher tower they reached the spacecraft level over 300 feet up to complete the first part of their journey to the moon. In the lift they passed a team of technicians dealing with a leaky valve in the liquid hydrogen piping system of the launch tower. This was quickly cured and, in any case, the particular part of the system was immediately by-passed, an operation made possible by NASA engineers' wise insistence on practically everything being in duplicate. By T minus 2 hours 40 minutes the 'close-out' team were helping the crew aboard the command module: spacecraft commander Armstrong into the left-hand couch, command module pilot Collins into the right-hand couch, and lunar module pilot Aldrin into the centre. By now the crew, the 'close-out' team and a fire-fighting rescue team in special heat-proof armoured cars were the only personnel within three miles of the Saturn. At this stage the emergency exit was: down the lift and into a 200-foot tunnel slide giving access to the blast escape room, a bunker embedded forty feet below the surface of the pad. Behind its six-inch steel door the simultaneous ignition of all the Saturn's fuel load, equivalent to a not-so-small nuclear explosion, would be noticeable only as a dull rumble from the sand-and-rubber-lined shelter mounted on twenty-four springs. This escape route would take about three minutes. In the event of that being too long, a more risky slide along a cable would take the crew to a point on the ground where, hopefully, the fire-fighting crew would pick them up and speed them away to safety.

It was developing into a sweltering Florida day as the crew began going through the last routines of the countdown; with the wind at ten knots and the thin cloud level at well above the tolerable limit, it was perfect launch weather. At T minus 43 minutes the Apollo access arm on the launch tower swung back to the stand-by position and the launch escape tower was 'armed' for possible use. All was ready for the final moments of the launch liturgy to be intoned by Jack King, the 'voice' of launch control:

This is Apollo 11 launch control. We have passed the six-minute mark in our countdown for Apollo 11; now five minutes, fifty-two seconds and counting. We are on time at the present time for our planned lift-off at thirty-two minutes past the hour. Spacecraft test conductor now has completed the status check of his personnel in the control room: all report that they are go for the mission and this has been reported to the test supervisor. The test supervisor is now going through some status checks. Launch operations manager reports go for launch. Launch director now gives the go. We are five minutes, twenty seconds and counting.

T minus 5 minutes Apollo access arm of the mobile launcher was now fully retracted. In any emergency the launch escape system (LES), the rocket attached to the top of the command module, would have been used to pull it and the crew clear of the launch pad.

Four minutes, fifteen seconds. The test supervisor now has informed the launch vehicle test conductor: 'You are go for launch.' ... Mark, four minutes and counting, we are go for Apollo 11.... Three minutes, twenty-five seconds and counting.... Neil Armstrong reported back when he received the good wishes [of the launch team]: *'Thank you very much, we know it will be a good flight.'* ... Firing command coming in now ...

T minus 3 minutes 10 seconds The firing command button was pushed at launch control centre (LCC), starting the fully automated computerized sequence of several hundred events leading up to lift-off.

T minus two minutes, forty-five seconds and counting ...

The launch team were now monitoring the 'red line values', vital temperatures and pressures in the rocket systems.

T minus one minute, fifty-four seconds and counting. Our status board indicates that the oxidizer tanks [of liquid oxygen] of the second and third stages have now pressurized [those of the first stage were already pressurized by this time]. We continue to build up pressure in all three stages in preparation for lift-off. T minus one minute, thirty-five seconds ... one minute, twenty-five seconds and counting. Our status board indicates that the third stage is completely pressurized.... Eighty-second mark has now been passed.... Fifty-five seconds and counting. Neil Armstrong has reported back: *'It's been a real smooth countdown.'*

T minus 50 seconds The launch vehicle was now switched onto its own internal electrical power sources. Four of the nine service arms of the mobile launcher tower were now retracted.

Power transfer is complete. We are on internal power with the launch vehicle at this time. ... All the second-stage tanks now pressurized. Thirty-five seconds and counting. We are still go with Apollo 11.... Astronauts report: '*It feels good.*' T minus twenty-five seconds ... twenty seconds and counting.

At T minus 17 seconds the guidance computer aboard the instrument unit went over to internal power.

T minus fifteen seconds ... twelve, eleven ...

T minus 10 seconds Positioned on either side of the flame trench below the Saturn, nozzles of the water deluge system began pouring water into it at a rate of 8,000 gallons per minute.

... nine ...

T minus 8.9 seconds Automatic ignition of the five F I engines of the S-IC first stage. Five plumes of flame began to roar down into the trench from the eighteen-foot-high, fourteen-foot-wide engine nozzles, vapourizing the water pouring from the deluge system.

... Ignition sequence starts. Six, five, four, three ...

T minus 2 seconds All engines were now running at ninety per cent of the seven and a half million pounds of thrust, gulping kerosene and liquid oxygen at the rate of 10,000 pounds per second.

... one ...

For the first time ever the normally monotone voice of Jack King at launch control began to break with emotion.

T minus 0 seconds Ground elapsed time, **GET 00:00:00** (hours:minutes:seconds) 2.32 p.m. BST. The 'launch commit' signal was flashed by computer to hold-down arms on the platform of the mobile launcher which gently released the 3,000-ton rocket, now some 86,000 pounds lighter since ignition. At the same moment the remaining service arms swung clear and the deluge system poured water down on the platform and into the flame trench at the rate of 50,000 gallons per minute.

. . . zero, all engines running. Lift-off, we have a lift-off.

The sound of the shouts and cheers of 3,500 newsmen was heard for a moment before it was drowned by the earthquake and thunder of the Saturn.

GET 00:00:02 A yaw manoeuvre, achieved by gimballing the four outer FI engines of the cluster of five, gently tilted the Saturn; eight seconds later the Saturn had cleared the launch tower, watched by close on a million people who had come to Cape Kennedy, including Hermann Oberth (the German rocket pioneer who had inspired von Braun) and many of America's political leaders.

GET 00:00:15 A roll and pitch manoeuvre pointed the Saturn out over the Atlantic. *'We've got a roll programme,'* reported Armstrong from the command module. *'Roll complete, attitude just perfect,'* he added a few seconds later as Apollo 11 headed out over the wide beaches of the Cape where children had scratched giant messages of good luck in the sand. Their destination, the moon, was now exactly 218,096 nautical miles distant across space as the majestic Saturn growled its way through scattered cloud high above the Cape and over the West Indies, where eight Russian naval ships on their way to Cuba enjoyed a grandstand view.

GET 00:00:51 The command module cabin pressure began to decrease as the rocket reached an altitude of 1,400 feet. Every half second the computer within the instrument unit reported on 1,348 systems within the Saturn.

GET 00:01:21 The moment of maximum dynamic pressure as the Saturn thrust its way through the atmosphere. Altitude now 43,365 feet, range from the launch pad 3·1 miles, speed 1,793 miles an hour.

GET 00:02:00 Armstrong reported go for staging. Fifteen seconds later he reported again as the centre of the five first-stage S-IC engines cut off. Altitude 145,600 feet, range 28·6 miles, speed 4,423 miles an hour.

GET 00:02:41 The four outboard engines of the first stage cut off. Altitude 217,655 feet, range 57 miles, speed 6,141 miles an hour. Between them the five first-stage engines had now burned up over 2,000 tons of fuel, reducing the total weight of the rocket by two-thirds. A second after the shutdown of the S-IC engines, eight solid-fuelled retro-rockets producing a thrust of 87,900 pounds each for 0·6 seconds slowed down the first stage, now separated from the S-II second stage by automatically triggered spring-loaded charges. At the same time four 21,000-pound thrust 'ullage' rockets mounted on the S-IC/S-II interstage were fired to settle the fuel in the bottom of the S-II tanks. A second later the five J 2 engines of the second stage began to build up thrust. Once off the launch pad, Apollo 11's journey became the responsibility of the flight controllers at Mission Control Centre (MCC) in Houston. '*Houston [this is] launch control: all engines are looking good,*' reported Cape Kennedy to MCC as the S-II achieved full power.

GET 00:03:11 In the command module cabin a light went out indicating that the S-IC/S-II interstage had been jettisoned as the second stage pushed the craft to an altitude of 301,266 feet. Range from the launch pad was now 100 miles, speed 6,439 miles an hour. Six seconds later the launch escape tower and boost protective cover were jettisoned. Armstrong: '*Tower.*' MCC: '*Roger, tower.*' In any emergency the command module would from now on have had to complete a space hop before returning to earth. The crew were enjoying the view: '*Houston, we advise the visual is go today.*'

GET 00:04:00 MCC reported to the command module that the trajectory and guidance looked good. At one-minute intervals from now until GET 00:08:00 Armstrong reported go from his end. MCC: '*Eleven, Houston, you are go at four minutes. ... Roger, Eleven, you're go from the ground at six minutes. ... Apollo Eleven, this is Houston, level flight time at eight plus one seven* [00:08:17; timings are in fact quite often given down to tenths, even hundredths, of a second. The launch was all of 724 thousandths of a second late!] ... *Outboard cutoff at nine plus one one* [00:09:11].' Armstrong: '*Apollo Eleven, go at seven minutes.*' MCC: '*Eleven, this is Houston, roger, you are go from the ground at seven minutes.*'

GET 00:07:40 Altitude 588,152 feet, range 690 miles, speed 12,778 miles an hour as the centre J 2 engine of the second stage

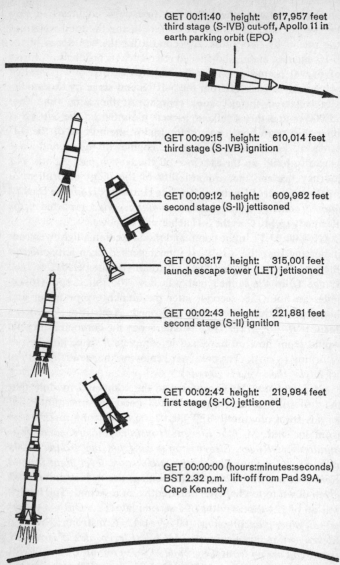

GET 00:11:40   height: 617,957 feet
third stage (S-IVB) cut-off, Apollo 11 in
earth parking orbit (EPO)

GET 00:09:15   height: 610,014 feet
third stage (S-IVB) ignition

GET 00:09:12   height: 609,982 feet
second stage (S-II) jettisoned

GET 00:03:17   height: 315,001 feet
launch escape tower (LET) jettisoned

GET 00:02:43   height: 221,881 feet
second stage (S-II) ignition

GET 00:02:42   height: 219,984 feet
first stage (S-IC) jettisoned

GET 00:00:00 (hours:minutes:seconds)
BST 2.32 p.m. lift-off from Pad 39A,
Cape Kennedy

Launch of Apollo 11 on 16 July 1969 into earth parking orbit on the way to the moon

shut down. MCC: '*Apollo Eleven, Houston, you're go at eight minutes . . . Apollo Eleven, this is Houston, you are go for staging . . . stand by for mode four capability.*' Mode four refers to the action the crew would have taken had anything gone wrong with the mission at this point. In that event they would have used the engine of the service module, having separated from the S-IVB and the lunar module to put them into orbit. The mission would then have been 'aborted' and they would have returned to earth. Armstrong: '*To mode four.*' MCC: '*Mark, mode four capability.*'

GET 00:09:11 Altitude 609,759 feet, range 1,018 miles, speed 15,468 miles an hour. Shutdown of the remaining four J 2 engines of the S-II second stage. The Saturn's fuel load was now lighter by about 964,000 pounds. A second later the S-II separated by four retro-rockets at the top of the second stage. After three seconds the single J 2 engine of the S-IVB third stage started up. Exactly ten minutes after lift-off from Cape Kennedy and perched high above the Atlantic, Armstrong reported that the Apollo craft was go for orbit, while from the ground came the computer-predicted time for the cut-off of the S-IVB engine. Armstrong: '*We're go.*' MCC: '*Apollo Eleven, Houston, you're go at eleven.*'

GET 00:11:42 Altitude 617,957 feet, range 1,639 miles, speed 17,380 miles an hour. The J 2 engine of the S-IVB shut down, and ten seconds later the Apollo 11 spacecraft was in orbit. The total weight of the craft and the S-IVB was now down to 299,586 pounds (134 tons), a mere four per cent of its launch pad weight. Armstrong: '*Shutdown, one-o-one point four by one-o-three point six.*' MCC: '*Roger, we copy shutdown, one-o-one point four by one-o-three point six.*' These figures referred to the parameters of their almost circular 101·4 by 103·6 nautical-mile earth parking orbit (EPO).

The Cape Kennedy team had truly come of age with the Apollo 11 launch, America's twenty-first manned space shot. The elegance and precision with which the 3,000 tons of the Saturn V were placed in orbit was an event totally unmatched in the history of technology and craftsmanship. Michelangelo would have admired the way in which so much brute force was moulded into such a fine moment. Someone monitoring

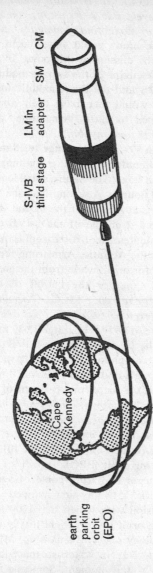

GET 02:50 (hours:minutes), speed 24,166 miles an hour, distance from earth: 2,090 miles, Translunar injection (T L I) prior to transposition and docking.

his console at launch control as the Saturn followed the predicted track perfectly, wondered whether it was really happening. 'It must be the computer simulation tapes,' he joked. '*It was beautiful. It was all a good ride*,' reported Armstrong from 100 miles above Africa. To the ever vigilant US Aerospace Defense Command, constantly on the watch for hostile missiles, Apollo 11 was simply recorded as 'Man-made Object in Space No. 4039'.

Safely in orbit, the crew removed their helmets and gloves as they, the flight directors in Houston, and the computers on board the CM and at MCC, started an exhaustive series of checks that was to last one and a half orbits, before the S-IVB engine was restarted to put them on the flight path to the moon.

GET 00:16 (hours:minutes) The tracking station in the Canary Islands took over from the dish at Cape Kennedy. In turn, the Canary Islands handed over to Tananarive in Madagascar, then Carnarvon in western Australia, the Honeysuckle dish near Canberra, and so on round the globe as they completed their ninety-minute orbit. Coming up over the complex of dishes at Goldstone in California, the Apollo 11 crew switched on their television camera for the first time but equipment on the ground was not yet ready. 'Not too unusual,' remarked some rather unkind person at MCC.

GET 02:30 Halfway round on a second orbit, 100 miles above the Carnarvon earth station, MCC gave Armstrong '*go for TLI*', the translunar injection burn of the S-IVB. Back in their helmets and gloves, the crew prepared for the restarting of the third-stage engine.

GET 02:44 (5.16 p.m. BST) Ignition at the start of the 5 minute 20 second burn of the J 2 engine which gradually increased Apollo's speed to 24,182 miles an hour, lifting it out of earth orbit towards the moon. As the engine cut off, the craft was already 200 miles along its quarter-million-mile flight path. A few moments later they passed through the Van Allen belts, but the dose of radiation they received was rather less than the dose a dentist uses to take an X-ray.

GET 02:56 The crew now altered the seating arrangements in the command module prior to the transposition and docking manoeuvre. Armstrong was now seated in the centre couch,

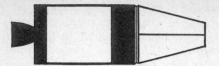

1 GET 03:15, CSM separation

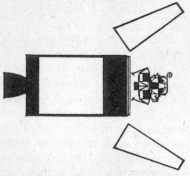

2 LM adapter panels jettisoned, CSM turns round

3 CSM approaches LM

4 GET 03:25, CSM docks with LM

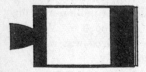
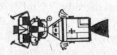

5 GET 04:12, CSM-LM separates from S-IVB

GET 03:15, transposition and docking manoeuvre

Collins in the left and Aldrin in the right. This arrangement was kept for the rest of the mission. The crew, still in their spacesuits, now pressurized the cabin to 5·7 pounds per square inch (PSI).

GET 03:15 With a sixteen-millimetre camera running, they separated the CSM from the S-IVB and the lunar module adapter (SLA) and, using the backward-pointing attitude control thrusters on the SM, drew a little in front of the Saturn. Again using the attitude thrusters, they turned the CSM around and docked the top of the CM with the lunar module docking hatch. By now the panels of the LM adapter had been jettisoned. This part of the manoeuvre completed, they released the catches holding the LM to the S-IVB and jetted themselves clear.

GET 04:40 The Apollo craft and the S-IVB were flying separately on the same flight path to the moon, but the Saturn third stage was not needed again and so the crew restarted its J 2 engine for the third time for a 2·8 second burn which slung the rocket into the beginning of an orbit round the sun. This is called the 'slingshot manoeuvre'.

GET 05:15 The crew were now able to relax and remove their helmets, gloves and the outer layer – the cumbersome pressure garment – of their spacesuits. Various checks were then carried out and the spacecraft batteries were connected to the fuel cells for charging. There followed an hour of star sightings for navigation purposes. At GET 07:00 (9.32 p.m. BST) the astronauts had their first meal in space. The menu: beef and potatoes, butterscotch pudding, four brownies and, to drink, grape punch. The crew were now settled for the long trip out to their lunar destination.

At GET 11:44 (2.16 a.m. BST, 17 July), according to their flight plan, the Apollo 11 crew were due to fire the main engine of the service module for a few seconds to carry out the first of four mid-course corrections (MCC-1) during the translunar coast (TLC). Coming nine hours after the translunar injection burn (TLI+9), this manoeuvre was designed to adjust the Apollo's speed and direction towards the appointed place in space behind the moon where the retrograde burn to put the craft into lunar orbit was to be made some sixty-four hours later. But in the event, computers at Mission Control and aboard the command module advised that the speed and direction of the TLI burn had been so precise that this first correction would not be necessary; any adjustments could wait for MCC-2 due at GET 26:44 (TLI+24, 5.16 p.m. BST).

During the translunar coast the Apollo was kept in what is called the passive thermal control (PTC) or 'barbecue' mode in which the spacecraft is automatically rolled gently at a rate of 0·3 degrees per minute as it glides towards the moon. The object of this is to prevent the sun from constantly heating one side of the craft while the other remains exposed to the coldness of outer space. Rotating the craft about its x-axis (an imaginary spit) causes it to be evenly exposed to the heat of the sun. PTC is initiated by thrusts from the attitude control system roll thrusters around the periphery of the service module.

GET 12:30 (3.02 a.m. BST, 17 July).For the crew it was now supper time. The menu this time: salmon salad, chicken and rice, six sugar cookie cubes, cocoa and a pineapple-grapefruit drink. Over the voice link the crew complimented the chef on his salmon salad, a generous gesture since like most of the food

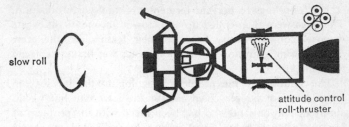

slow roll

attitude control
roll-thruster

Passive thermal control (P T C)

aboard it was freeze-dried. The Apollo 11 larder consisted of some seventy different rehydratable freeze-dried 'dishes'. Each complete meal consisted of four or five different selections, and a daily ration pack contained three of these meals. In addition to the daily meal packs for each astronaut, food for five extra days was carried in a 'snack pantry'. This was not packed into separate meals, so that a member of the crew wanting to have a nibble would not have to break into a regular meal pack somewhere deep down in the storage box. The language of space has a word for everything and these between-meal snacks were packages in what is called the 'smorgasbord mode'.

Water for drinking and for rehydrating the food was obtained from three sources in the command module: a dispenser for drinking water and two water spigots, one hot and one cold. The drinking water dispenser produced a continuous flow operated by a trigger, while the others were designed to squirt water in one-ounce units into the plastic freeze-dried food packets. Some of the water used was supplied from the service module where hydrogen and oxygen were reacted together in fuel cells to produce electric power and water. After he had injected water from the hot or cold spigot into the food pack, the astronaut-cook for the day would knead the packet for three minutes. To eat the contents, the neck of the packet was cut off and mushy food was squeezed into the mouth. Such are the refinements of the *haute cuisine* of weightlessness.

After each meal the astronauts put germicide pills into the empty food bags to prevent fermentation and gas production. The bags were then rolled up and stowed in one of the waste disposal compartments. To keep their teeth clean the crew used chewing gum and edible toothpaste; wet tissues replaced table napkins.

The solution to the problem of going to the lavatory in space may not be very pleasant for the astronaut but would surely delight anyone who has been faced with the problem of looking after a baby. Solid wastes are collected in plastic defaecation bags which contain a germicide to prevent bacterial activity. These are stowed for post-flight analysis. Urine is likewise collected in bags and is dumped out into space through valves in the command module. During the Gemini flights, crews would save this event for sunset or sunrise and watch the 'fireworks' display as the globules of urine glinted in the sunlight. At first the medical staff at Mission Control were puzzled by the way in which astronauts seemed to produce urine at exactly ninety-minute intervals. During extravehicular activity (EVA) the innermost layer of the astronaut's spacesuit is nothing more complicated than a nappy, though he can still get rid of urine through a tube into a reservoir he carries.

Keeping the body clean is very much a problem, and damp towels are provided for this purpose. Shaving turned out to be quite easy. The Americans spent thousands of dollars designing an electric razor with a vacuum cleaner attached to prevent the spacecraft from becoming filled with weightless whiskers, but, as the Apollo 10 crew demonstrated, shaving foam and a safety razor work just as well, the whiskers remaining in the sticky lather.

After the first 'evening' meal on the Apollo 11 mission, the crew made a second more successful attempt to send colour television pictures back to earth. MCC: '*Apollo Eleven, Houston, we'll try once more on this TV request, we would like ten minutes' worth of TV, and we would like a narrative if you can give us one on the exterior shot. We also suggest you might try an interior position, over.*' By now the blue planet earth with its white, swirling cloud cover was on the giant screen at Houston. Armstrong: '... *we can clearly see the western coast of North America ... Baja California and Mexico down as far as Acapulco and the Yucatan peninsula and we can see all through Central America to the northern coast of South America: Venezuela and Columbia. I'm not sure you will be able to see that on your screen down there.*' Released in America two hours later, the pictures were described as the clearest yet, but viewers in Europe were not so lucky, perhaps because of the breakdown of an Atlantic satellite a few weeks before Apollo 11 was launched, which meant that the pictures had to make two hops from Houston across the Pacific and Indian Oceans – a journey of over 100,000 miles, nearly twice the distance between the Apollo 11 spacecraft and earth.

GET 13:32 (4.04 a.m. BST, 17 July) After going through a 'pre-sleep checklist', which included a report from Armstrong to MCC on the 'crew status', the Apollo astronauts settled down for a nine-hour rest period. The crew status report, which was made twice a day throughout the flight, was to tell the medical team at MCC how much radiation each member of the crew had been exposed to and whether they had taken any medicine. The command module medical kit to the right of Aldrin's couch contained three motion sickness injectors, three pain suppression injectors, one two-ounce bottle of first aid ointment, two two-ounce bottles of eyedrops, three nasal sprays, two compress bandages, twelve adhesive bandages, and a thermometer. Pills carried on the mission included sixty antibiotic, twelve anti-nausea, eighteen pain killing, twelve stimulant, sixty decongestant, twenty-four anti-diarrhoea, seventy-two aspirin and twenty-one sleeping. A similar, smaller kit was carried in the lunar module.

Keeping the Apollo 11 crew fit during the mission was the responsibility of Dr Berry and his team at MCC. To help him do this each member of the crew was 'wired up' with biomedical monitoring pads attached to their bodies which recorded such factors as heart and breathing rates. At lift-off the crew's heart rates were: Armstrong, 110; Adrin, 88; Collins, 99 (the normal rate for the human heart is seventy beats per minute). Information from the monitoring pads was routed automatically through the spacecraft's communications system, back down to the manned spaceflight network's (MSFN) tracking stations on earth, and into NASA's two-million-mile communication system (NASCOM) to reach MCC in Houston some fractions of a second later.

On Apollo missions all communication between the spacecraft and earth – the telemetry, tracking, automatic commands and voice circuits – are combined in a single radio carrier wave which involves the use of only one antenna. This is called the 'unified S-Band system' and involves a staff of 4,500 distributed among fifteen tracking stations and numerous switching centres around the globe.

While the spacecraft is in earth parking orbit and just before splashdown, communications are routed through thirty-foot dishes at eleven sites on land and aboard four tracking ships. In addition to these thirty-foot dishes, eight Apollo range instrumentation aircraft (ARIA) operate in the Atlantic, Pacific and Indian Oceans. These are modified Boeing jets which carry seven-foot antennae in their bulbous noses. Once the spacecraft has completed the translunar injection burn and is on its way out to the moon, three more powerful eighty-foot dishes – one at Goldstone in California, one near Madrid, and one at Honeysuckle Creek near Canberra – come into action. These deep-space tracking stations are so placed that, while the earth revolves on its daily axis, one of them can always 'see' the moon. On the Apollo 11 mission two even more enormous 'electronic ears', the 210-foot Mars antenna at Goldstone and the 210-foot dish at Parkes in Australia, were used. The Mars antenna was brought in to assist communications with the lunar module once it had separated from the command module in lunar orbit, while the Parkes dish was used to receive the television transmissions from the surface of the moon.

Apart from handling the voice, telemetry and command circuits between the spacecraft and earth, the dishes are also used to track the craft in space. By measuring the round-trip time of a ranging signal, automatically re-transmitted by the spacecraft and measuring the Doppler shift (the re-transmitted signal's frequency change, like the change in note of the whistle of a passing railway train), it is possible to determine the range of the craft to within a few hundred feet and its range rate or speed to within a few inches per second.

Linking all the tracking dishes of the MSFN is NASA's staggering global communications network (NASCOM). Using a web of about two million miles of cable, microwave, radio and satellites link, it is centred at the Goddard Spaceflight Center (GSFC) in Maryland. Information originating from the Apollo 11 spacecraft ('down-link data') such as the heart and breathing rates of the astronauts, the command module-cabin pressure and temperature, etc., was transmitted automatically, second by second ('real-time' telemetry), through the cluster of four small S-Band dishes deployed outside the service module at the rate of 51,200 bits per second. Picked up by a ground station, it was then processed by computer before being passed on to Houston at a modest 2,400 bits per second. These computers at the tracking sites are also used for relaying commands to the spacecraft's systems for passing information ('up-link data') to the spacecraft's own computers at the rate of 1,200 bits per second. Several times a second the computers at Houston 'talk' to the other computers in the network or in the spacecraft. The MCC computers compare everything they receive from other computers with their memories and then indicate problem areas and other relevant data to the flight controllers. Another important function of the Houston computers is time-keeping, and they deal in thousandths of rather than whole numbers of seconds. Between Houston and the centre of the NASCOM web at the Goddard Spaceflight Center there are two 50,000-bit-per-second circuits. The volume of traffic handled at Goddard during an Apollo mission is equivalent to a 500-page novel per second. The computers there would take less than a second to add a column of ten-digit numbers three-quarters of a mile high. To talk to their operators, the computers around the network are capable of slowing down to a more reasonable 100 human language words per minute.

While Goddard is the switching centre for communications, it is at Mission Control, Houston, that all important decisions are taken during an Apollo flight. During the Apollo 11 mission a team of flight directors under the director of flight operations,

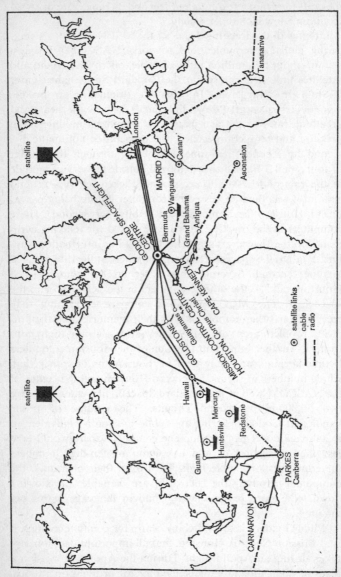

The main elements of NASA's two-million-mile Apollo communications network (NASCOM)

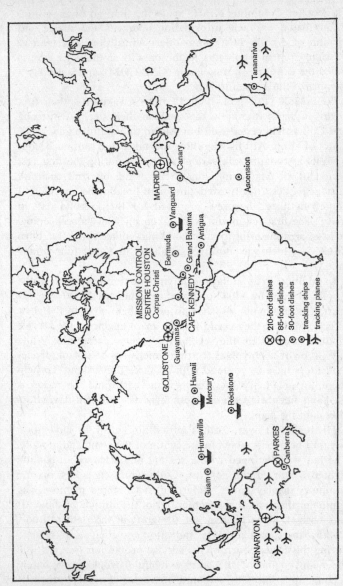

Apollo manned spaceflight tracking network (M S F N)

Tananarive

Canary

MADRID ⊕

Ascension

Vanguard ◢

Bermuda
Grand Bahama
Antigua

MISSION CONTROL
CENTRE HOUSTON
Corpus Christi

CAPE KENNEDY ◉

GOLDSTONE ⊗
Guayamas ⊕

⊗ 210-foot dishes
⊕ 85-foot dishes
◉ 30-foot dishes
◢ tracking ships
✈ tracking planes

◉ Hawaii
Mercury
◉ Redstone

Huntsville
◉

Guam ◉

CARNARVON

PARKES ⊗
Canberra

Christopher Columbus Kraft, made the final go or no-go decisions based on their information and with the advice and opinion of Apollo 11's commander, Armstrong. The man at Mission Control who talks to the crew over the voice link is called the capsule-communicator (CAP-COM), and is usually a member of the astronaut team.

GET 22:30 (1.02 p.m. BST, 17 July) Awake after their first 'night' in space, the crew reported on their sleep. Armstrong and Collins had rested well but Aldrin only got five and a half hours of sleep. At this stage this did not cause too much concern, as astronauts rarely slept very well during the first rest period of an Apollo mission. Breakfasting off fruit cocktail, sausage patties, four toasted-cinnamon bread cubes, cocoa and grapefruit juice, the crew prepared for their second day in space. The first few hours were taken up with 'housekeeping' and star sighting prior to the second mid-course correction burn (MCC-2) which was now to go ahead in about four hours' time.

In addition to assimilating the information passed up to the Apollo 11 crew by MCC on the range and speed of the spacecraft, the astronauts also had to point themselves in the right direction before they could fire the main engine of the service module to complete the second mid-course correction. While navigation at sea involves the two dimensions of a plane, space-navigators have to contend with all three dimensions. To help them, an imaginary line between the earth and the moon is used and, like their seafaring counterparts, astronaut-navigators also sight the stars.

His breakfast over, Collins settled himself in the navigation bay, just below his couch in the centre of the command module. The bay was equipped with a sextant and a telescope. Having sighted a star in the sextant, Collins pressed a key on the computer display panel. Referring to its stored memory, the computer then told him whether or not the attitude of the craft was correct. Once he had got the craft in the right attitude, Collins reset the attitude indicators so that any change during the critical burn of the service propulsion system (SPS, the main engine of the service module) would immediately become apparent. The onboard computer also keeps track of attitude using not the stars, but electronics associated with the imaginary line between the earth and the moon.

GET 24:39 (3.11 p.m. BST, 17 July) Collins: *'Houston, Apollo Eleven, on this star auto manoeuvre worked just fine. I'm right at the substellar point. Everything looks beautiful except there's no star in sight. It's just not visible.'* MCC: *'Roger, this is for star 01?'* Collins: *'That's correct.'* MCC: *'You're not getting any reflections, or anything like that, that would obscure your vision, are you?'* Collins: *'Well, of course the earth is pretty bright, and the black sky instead of being black has a sort of rosy glow to it and the star, unless it's a very bright one, is probably lost somewhere in that glow, but it just is not visible.'* MCC: *'Roger, we copy. Stand by a minute, please. . . . Eleven, this is Houston, can you read us the shaft and trunnion angle off the counters?'* Collins: *'I'll be glad to. Shaft: 331·2, and trunnion: 35·85.'* MCC: *'Roger, thank you.'*

Collins: *'It's really a fantastic sight. Through that sextant a minute ago, during that auto manoeuvre, the radical swept across the Mediterranean. You could see all of North Africa, absolutely clear, all of Portugal, Spain, southern France, all of Italy, absolutely clear. Just a beautiful sight.'* MCC: *'Roger, we all envy you the view up there.'* Collins: *'But still no star.'* When Collins had sorted the problem out, all was ready for the mid-course correction.

GET 26:46 (5.18 p.m. BST, 17 July)

This is Apollo control at twenty-six hours and forty-six minutes. Apollo Eleven's distance from the earth is 109,245 nautical miles [125,632 statute miles], its velocity is 5,033 feet per second. Spacecraft weight: 96,361 pounds . . . the duration [of the MCC-2 burn] will be three seconds . . . burning . . . shutdown . . .

MCC: *'Looks like the SPS is definitely go.'* Armstrong: *'Good to hear it.'*

The SPS engine developed a thrust of 21,500 pounds by burning a mixture of two hydrazine compounds, described as a hypergolic propellant since it ignited spontaneously on contact with the oxidizer, nitrogen tetroxide in this case. It behaved perfectly, the first time it had been swiched on, for MCC-2. It has to be restartable because it can be used as many as eleven or more times during an Apollo mission, not only for mid-course corrections but also for getting the craft into lunar orbit and for the vital burn to bring it back to earth.

The MCC-2 burn over, the crew relaxed as there was little to do until the next rest period except eat. By now the Apollo 11 craft was over halfway to the moon but its speed had dropped from 24,182 miles an hour at TLI to a mere 3,422 miles an hour due to the pull of earth's gravity. It was still some time before the pull of the moon began to speed them up again. At this point in the mission the crew gave us earthlings their first really detailed descriptions of the view through the command module windows.

GET 27:14 (5.46 p.m. BST) Aldrin: '*Hey, Jim* [Jim Lovell of the Apollo 8 crew, who was acting as capsule-communicator (CAP-COM) at this time], '*I'm working through the monocular* [telescope] *now and the view is out of this world. I can see all the islands in the Mediterranean from larger and smaller islands of Majorca, Sardinia, Corsica, a little haze over the upper Italian peninsula, a few cumulus clouds out over Greece, sun setting on the eastern Mediterranean now. . . . The British Isles are definitely a greener colour than the brownish-green that we have in the Mediterranean islands and in Spain.*' MCC: '*Roger. I understand that the northern African-Mediterranean area is clear today.*' Aldrin: '*Right, we see a bunch of roads with cars driving up and down too.*' MCC: '*Do you find the monocular is any good to you, Buzz?*' Aldrin: '*Yeah, it would be nice if it had another order of magnif power on it. It has a tendency to jigger around a little bit . . .*'

Aldrin: '*Houston, Apollo Eleven.*' MCC (CAP-COM, Lovell): '*Go ahead, Eleven.*' Aldrin: '*I got a comment about the point on the earth where the sun's rays reflect back towards us. In general, the colour of the ocean is mostly uniform and it's bright, darker blue, except for that region that's about one-eighth of an earth's radius in diameter. In this circular area, the blue of the water turns a greyish colour and I'm sure that's where the sun's rays are being reflected back on up toward us. Over.*' MCC: '*Roger, Buzz. We* [the Apollo 8 crew] *noticed the same thing. It's very similar to looking at a light shining on* something like a billiard ball, or a bowling ball. You get this bright spot and the blue of the water then turns into a sort of greyish colour.' Aldrin: 'Yeah. Is there a Navy term for that?' MCC: 'A lot of grey paint.'

Aldrin: 'Looks like the best way to get a good steady view through the monocular is just to steady it out and let it float right in front of your eye, and then you kind of float up next to it so that you're not touching it at all. It has a very slow drift and you get a better steadier view that way.' MCC: 'Sounds good. How does it feel to be airborne again, Buzz?' Aldrin: 'Well, I'll tell you. I've been having a ball floating around inside here, back and forth, from one place and back to another. Just like being outside [space-walking], except more comfortable.' MCC: 'A lot bigger than the last vehicle.' (Aldrin's last space trip was in the comparatively cramped surroundings of the Gemini 12 capsule, which he shared with Jim Lovell.) Aldrin: 'Sure, it's nice in here. I've been very busy so far. I'm looking towards taking the afternoon off. I've been cooking and sweeping and almost sewing and, well, you know, the usual housekeeping things.'

GET 28:07 (6.39 p.m. BST) Collins: 'I've got the world in my window for a change and looking at it through the monocular, it's really something. I wish I could describe it properly. The weather is very good, South America is coming around into view. I can see on what appears to me to be the upper horizon a point that must be just about Seattle, Washington, and from here I can see all the way down to the southern tip of Tierra del Fuego in the southern tip of the continent.' MCC: 'Roger, sounds like you've got a beautiful view up there.' Collins: 'Absolutely fantastic. I hope the pictures come out [taken with a still Hasselblad camera]. We're rotating around and it's going out of view again.' MCC: 'Sounds like one of those rotating restaurants.'

The astronauts were now eating their 'midday' meal: frankfurters with apple sauce, chocolate pudding and an orange-grapefruit drink.

GET 28:40 (7.12 p.m. BST) MCC: 'Is that music I hear in the background?' Apollo 11: 'Buzz is singing. . . . Pass me the sausage . . .'

GET 29:34 (8.06 p.m. BST) Collins: *'Houston, we're just looking at you out of the window here. Looks like there's a circulation of clouds that just moved east of Houston over the Gulf* [of Mexico] *and forward area. Did that have any rain, do you know, this morning?'* MCC: *'Roger, report from outside says it's raining out here and looks like you got a pretty good eye for the weather there.'* Collins: *'Well, it looks like it ought to clear up pretty soon from our viewpoint. The western edge of the weather isn't very far west of you. Hey, you got any medics down there watching heart rate? I'm trying to do some running in place down here* [in the navigation bay of the command module] *and I'm wondering, just out of curiosity, whether it brings my heart rate up.'* MCC: *'They will spring into action momentarily, stand by. . . . Hello, Eleven. We see your heart beat. . . . Eleven, Houston, Mike, we see about a ninety-six heart beat now.'*

GET 32:50 (11.22 p.m. BST) MCC: *'Can you pick out anything around Edwards* [Air Force Base], *the dry lake or anything? Over.'* Aldrin: *'Negative. I don't have that resolution, but to give you some idea I can, on the lower Texas coast I can see, knowing what I'm looking for, I can see Padre Island. I can just barely make out the fact that there is a thin bit of land, then there's a little dark zone which is the Laguna Madre between it and the mainland.'* MCC: *'Roger, that's pretty significant. Thank you very much, Buzz.'*

Aldrin: *'How far out are we?'* MCC: *'Stand by, I'll give it to you exactly. It looks like around 130,000* [nautical miles] *but stand by. . . . Exact range is 125,200* [nautical] *miles* [143,980 statute miles] *and you're travelling at 4,486 feet per second* [3,050 miles an hour].' Aldrin: *'Pretty far and pretty slow. Just past halfway. What's the latest now on Luna 15?'* MCC: *'Stand by, I'll get the straight story for you. . . . Hello, Apollo Eleven, Houston, over.'* Aldrin: *'Go ahead.'* MCC: *'Roger. Latest on Luna 15: Tass* [the Soviet news agency] *reported this morning that the spacecraft was placed in an orbit close to the lunar surface and everything seems to be functioning normally on the vehicle. Sir Bernard Lovell* [whose team at Jodrell Bank in England was using the giant 250-foot dish to track the Soviet probe] *says the craft appears to be in*

an orbit of about sixty-two nautical miles, over.' Aldrin:
'Okay, thank you.' MCC: *'Also, President Nixon is reported
to have declared a Day of Participation on Monday for all
Federal* [government] *employees to enable everybody to follow
your activities on the surface. So it looks like you are going
to have a pretty large audience for the EVA* [extravehicular
activity] *on the moon.'* Aldrin: *'Well, that's very nice. I'll tell
Neil* [Armstrong] *about it.'*

GET 34:00 (00.32 a.m. BST, 18 July) Before the last meal
of the day and a ten-hour rest period, the Apollo 11 crew
turned on the twelve-pound Westinghouse colour television
camera for a half-hour transmission to earth. The first pictures
included a close-up shot of one of their computers. MCC:
'We see the "diskey" [computer display console] *flashing with
a 651.'* Apollo 11: *'Aerial* [the four-dish S-Band antenna
attached to the SM] *high gain angle, telling us which way the
earth is. We also give you the time of day in our system of
mission elapsed time* [ground elapsed time, GET]: *thirty-four
hours, sixteen minutes and umpteen seconds. We see out our
side window the sun going by and, of course, out of one of
our windows right now we've got the earth.'* MCC: *'Eleven,
Houston, we see a box full of goodies there, over.'* Armstrong:
*'Would you believe you're looking at chicken stew here? All
you have to do is add three ounces of hot water for five or
ten minutes. We get our hot water out of a little spigot up
here . . . and we just stick the end of this little tube* [on the
plastic packet of freeze-dried chicken stew] *in the end of the
spigot and pull a trigger three times, for three ounces of hot
water, and then mush it up . . . and there you go, beautiful
chicken stew.'* Armstrong then gave a weightless demonstration
using a flashlight which he set rotating slowly in mid air.
The camera was then pointed out of one of the windows to
give a view of earth. The transmission over, the crew settled
in for the 'night', two of them in sleeping-bags below the row
of couches and one of them in the left-hand couch.

GET 48:00 (2.32 p.m. BST) Following a bonus one-hour
lie-in, the Apollo crew were eating breakfast at the start of
their third day in space. They had rested well: after only five
and a half hours the night before, Armstrong and Aldrin had

got eight hours' sleep, while Collins chalked up nine hours, a record 'space sleep' only equalled by the tired crew of Apollo 10 on their homeward journey from the moon in May. It was to be another comparatively restful day; Mission Control were so pleased with the trajectory that the third mid-course correction (MCC-3) originally scheduled for GET 53:55, twenty-two hours before they went into lunar orbit, was cancelled.

After lunch – cream of chicken soup, turkey and gravy, cheese cracker cubes, chocolate cubes, and a pineapple-grape-fruit drink – the crew prepared to crawl through the docking tunnel into the lunar module for the first time to see if it had come safely through launch. Before they did this they switched on the colour television camera.

GET 57:00 (11.30 p.m. BST) The crew were now preparing to pressurize the cabin of the lunar module. The oxygen pressures in the command module and the lunar module now had to be equalized as Armstrong and Aldrin prepared to open the hatch at the LM end of the tunnel and enter the lunar module for the first time. First the command module hatch was opened. Then they began to remove the docking equipment from the tunnel linking the two craft.

MCC: *'Eleven, Houston, that's a beautiful picture. We've got a twelve-second delay.'* The colour television signal, picked up by the Goldstone tracking station, was being converted at Mission Control to a line standard acceptable to the global television networks. This process involved a delay of twelve seconds. Apollo 11: *'Mike must have done a smooth job in that docking* [the docking during the transposition and docking manoeuvre just after translunar injection], *there isn't a mark on the* [docking] *probe.'* Armstrong and Aldrin were now removing the probe from the command module end of the tunnel. MCC: *'We're really getting a great picture here, Eleven, over. . . . Eleven, Houston, with a twelve-foot cable we estimate you should have about five to six feet excess when you get the camera into the LM, over. . . . We can see the drogue now.'* The drogue assembly is a cone-shaped structure fitted to the outside of the lunar module's docking hatch at the top of the LM's section of the thirty-two-inch diameter docking tunnel. It is

1. *Top left:* Apollo 11 commander Neil Armstrong.
2. *Top right:* Apollo 11 lunar-module pilot Edwin Aldrin.
3. *Bottom left:* Apollo 11 command-module pilot Michael Collins.
4. *Bottom right:* Wernher von Braun.

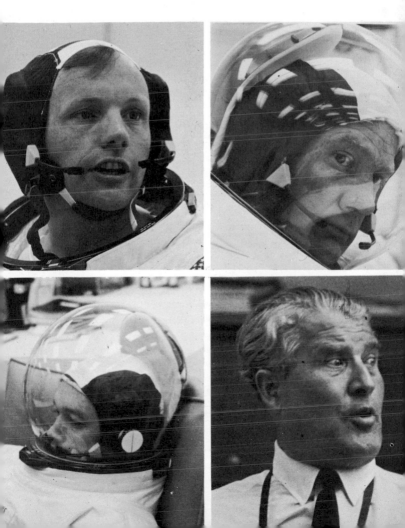

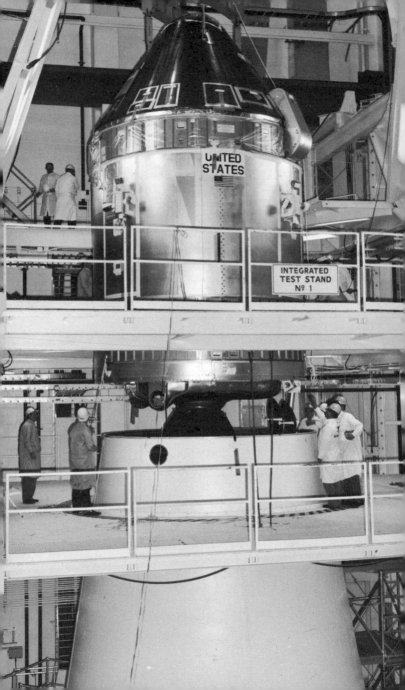

5. *Left:* The Apollo 11 command and service module is lowered atop the Saturn V stack in the VAB high-bay area.
6. *Below:* The Apollo 11 lunar module.

7. The crawler begins the journey to the moon. It carries the Apollo 11 Saturn V moonship and its mobile launcher on the first stage of the 3½-mile trip from the VAB to the launch pad.

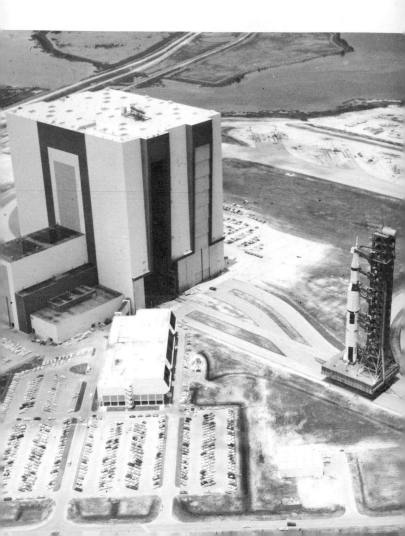

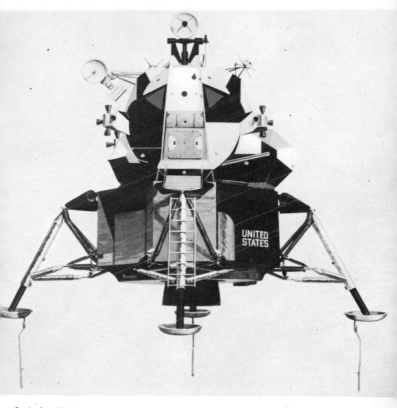

8. A detailed drawing of the lunar module with its legs fully extended. Attached to the foot pads are probes which tell the pilot when he is about to touch down on the moon. A ladder runs down from the ascent stage's exterior hatch. The descent engine can be seen under the craft behind the front leg. The two front windows are either side of and just above the hatch. Beyond these can be seen two sets of attitude-control thrusters at the two front corners of the craft. Above the hatch is the rendezvous radar dish antenna. To the left of this is the S-band dish antenna used for speech, data and television communication with the earth. Opposite this on the right of the radar dish is the VHF antenna used for communication with the CSM.

9. *Top:* The scene in the firing room at Launch Control Centre at Cape Kennedy. A similar room in Houston controls the mission once it has left the pad.

10. *Bottom:* A telescopic view of the first stage of a Saturn V being jettisoned at an altitude of 38 miles.

11. *Right:* The moon as seen by the departing Apollo 10 crew. The square marks the area in the Sea of Tranquillity where the Apollo 11 landing was made.

12. *Left:* Neil Armstrong inside the lunar module. Above his head is the hatch which leads to the command module.
13. *Below:* Edwin Aldrin using a core sampler to extract a sample from the Sea of Tranquillity.

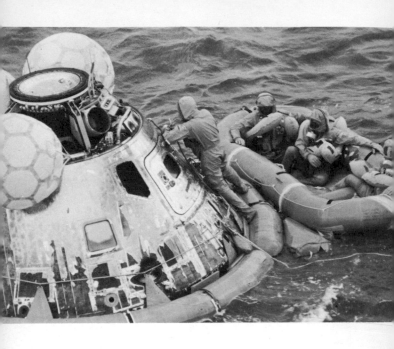

14. *Above:* Wearing biological isolation garments the three Apollo 11 astronauts watch a frogman using a decontaminant spray on the command module following splashdown.
15. *Right:* An Apollo 11 astronaut being winched aboard the recovery helicopter.

16.   *Above:* Apollo 8
commander Frank
Borman (*left*) and
NASA's administrator
Thomas Paine watch
the launch of Apollo
12 through a thunder
cloud.
17.   *Left:* Moon
sample No. 10003. The
first moonrock
brought back by the
Apollo 11 crew to be
examined in detail.

the 'female' counterpart of the command module's 'male' docking probe. This too had to be cleared from the tunnel to allow the two members of the crew to gain access to the lunar module.

MCC: *'Real goof view of the LM hatch handle there, Eleven, over.'* Armstrong: *'Looks like we'll be ready to go into the LM early, if that's OK with you all down there.'* MCC: *'Roger, it's fine with us, Neil, go ahead any time you wish, over.'* Armstrong: *'OK, the dump valve is actuated.'* This valve let the oxygen from the command module into the lunar module's cabin and so equalized the pressure in the two craft. *'Eleven, Houston, we're really amazed at the quality of the picture up in the tunnel, it's really superb, over.'* Armstrong: *'We're about to open the hatch now.'* Collins (who was operating the camera): *'When you open up the door . . . it turns the lights on. How about that, it's just like a refrigerator.'*

Aldrin and Armstrong then showed their audience briefly round the cabin of the lunar module. Aldrin pointed out the camera bracket on the right-hand triangular window to which he would attach the cine camera when filming Armstrong as he made man's first descent onto the lunar surface.

'Voice' of Mission Control, a 'public affairs' officer (PAO):

We've been receiving television now from the spacecraft for about an hour and twenty minutes. Apollo Eleven is presently 177,000 [nautical] miles [203,550 statute miles] from the earth.

Aldrin then held up the helmets which he and Armstrong would wear for their moon walk. They have two visors, in addition to the basic spacesuit helmet's visor; one is to protect the astronaut from possible showers of micrometeorites, the second, gold-coated, to protect the astronaut's eyes from the intense light of the sun in the vacuum of space. Armstrong: *'You did say this* [transmission] *is going out now, did you?'* MCC: *'Stand by, think so. Eleven, you've got a pretty big audience: it's live in the US, it's going live to Japan, Western Europe and much of South America. Everybody reports very good colour, they appreciate the great show.'* Armstrong: *'Roger, understand, thank you.'* The lunar module passed its first inspection and Armstrong and Aldrin returned to the command module for their supper.

During the day the Luna 15 drama entered a new phase following a telephone call from Frank Borman, the Apollo 8 commander, recently back from a trip to the Soviet Union, to the president of the Soviet Academy of Sciences, Dr Keldysh. The result was a telegram which affirmed that the Luna's orbit would not interfere with the Apollo 11 mission. Luna 15 remained an enigma.

GET 60:00 (2.32 a.m. BST, 19 July) The Apollo 11 crew bedded down at the end of their third day in space. The ninety-minute television transmission had delighted the three or four thousand press and television representatives who had now gathered in Houston. They had earlier been complaining to the public affairs office of NASA about the crew's reticence. Told at a press briefing that perhaps this crew were just not talkative types, one desperate Italian journalist had asked, hopefully: 'Do they talk in their sleep?' Receiving a negative answer, he muttered: 'That monkey at least dreamt.' He was referring to a monkey that was launched into orbit before the Apollo 11 mission for an intended month in space. It was brought down early and died almost immediately of undetermined causes.

GET 70:58 (1.30 p.m. BST, 19 July) MCC: '*Apollo Eleven, Apollo Eleven, this is Houston, over.*' Apollo 11: '*Good morning again, Houston.*' Two hours earlier the astronauts had reported to Mission Control at the end of their planned nine-hour rest period, but the flight was still going so well at this stage that the fourth mid-course correction (MCC-4) had been cancelled. The spacecraft was within two feet per second of its planned speed and no further correction to the translunar flight path would now be necessary, so Mission Control had told the crew to go back to sleep. Awake now, the astronauts were having their breakfast: Canadian bacon with apple sauce, sugar-coated cornflakes, peanut cubes, cocoa, and an orange-grapefruit drink. The crew had rested well and the medical team at Houston were pleased with their state of health. The day would end with the Apollo craft in lunar orbit at the beginning of the crucial sixty hours of the mission, and the astronauts were fresh and ready.

GET 71:50 (2.22 p.m. BST) Collins: '*Houston, Apollo Eleven.*

The earthshine coming through the window is so bright you can read a book by it.' Two days before, Collins had reported trouble trying to sight stars because of the brightness of the earth, but now, as they were drawing away from the planet, the darkness of outer space was beginning to descend.

GET 71:58 (2.30 p.m. BST) Apollo 11: '*Houston, it's been a real change for us. Now we're able to see stars again and recognize constellations for the first time on the trip. The sky is full of stars just like the night sky of the earth, but all the way here we've only been able to see stars occasionally.*' MCC: '*I guess it's turned into night up there.*' Apollo 11: '*It really has.*'

One of the 'household' chores which the astronauts carried out during the 'morning' of their fourth day in space was to change the cabin atmosphere filter. This consisted of a removable canister containing lithium hydroxide and activated charcoal which served to remove carbon dioxide and body odours from the spacecraft atmosphere of oxygen. They also 'purged' the fuel cells aboard the service module. These cells, a British invention, produce electricity and water from the chemical reaction of liquid hydrogen and liquid oxygen, and power the spacecraft from a few moments before lift-off until separation of the service module just before splashdown. From these cells came the power with which the spacecraft's emergency batteries were charged from time to time. At intervals during the flight the fuel cell units were purged, or flushed out with liquid hydrogen to clean out any impurities that might have accumulated.

GET 72:33 (3.05 p.m. BST) MCC: '*You might be interested in knowing, since you are already on the way, that a Houston astrologer, Ruby Graham, says that all the signs are right for your trip to the moon. She says that Neil is clever; Mike has good judgement and Buzz can work out intricate problems. She also says that Neil tends to see the world through rose-coloured glasses but he is always ready to help the afflicted or distressed. Neil, you are also supposed to have "intuition that enables you to interpret life with feeling". Buzz is supposed to be very sociable and cannot bear to be alone, in addition to having excellent critical ability. Since she didn't know at what hour Mike was born she has decided that he either has the same attributes as Neil or he is inventive with an unconventional attitude that*

91

might seem eccentric to the unimaginative. I checked with flight operations for all the signs for the mission.' Apollo 11: *'Thank you very much.'*

GET 73:16 (3.48 p.m. BST) MCC: *'Apollo Eleven, this is Houston. We're on low-bit rate* [referring to the speed of the automatic telemetry between spacecraft and ground tracking stations] *at the present time so it's going to take us a little longer than normal to get this stuff up to you.'* Apollo 11: *'I guess we're in no rush. The view of the moon we've been having recently is really spectacular. It fills about three-quarters of the hatch window and, of course, we can see the entire circumference even though part of it is in complete shadow and part of it is in earthshine. It's a view worth the price* [$350 million] *of a trip.'*

GET 75:32 (6.02 p.m. BST) The Apollo craft had now been coasting through space for three days and had covered over 200,000 miles. Accelerating under the influence of lunar gravity, the craft was approaching the moment when it would disappear behind the moon to make the critical burn of the service module engine that would brake the craft into its first egg-shaped lunar orbit. The engine was to be fired retrograde for six minutes, taking 1,988 miles an hour off the speed. If that burn had failed, the craft would, quite simply, have swung around the moon and headed back to earth; had it lasted too long, the craft would have crashed onto the moon. It was a tense moment in the flight, decreed by the laws of celestial mechanics to take place while the spacecraft was out of sight and out of contact, its crew of three alone in the silence of space. As Apollo 11 set below the lunar horizon, all Mission Control could do was to reassure the astronauts and then just wait. MCC: *'Apollo Eleven, this is Houston. All your systems are looking good going around the corner. We'll see you on the other side.'* Armstrong: *'Roger, everything looks OK up here.'*

GET 75:41:23 As the four dishes of the S-Band antenna on the service module were lost by the eighty-five-foot electronic ear at the Madrid station, the eternal primeval cosmic crackle replaced the link with the spacecraft.

PAO: 'We have loss of signal [LOS] as Apollo Eleven goes behind the moon. Velocity 7,664 feet per second [5,212 miles an hour],

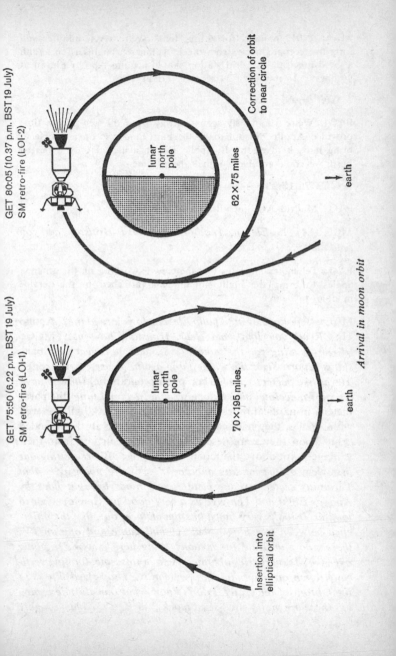

GET 75:50 (6.22 p.m. BST 19 July)
SM retro-fire (LOI-1)

lunar north pole

70×195 miles.

earth

Insertion into elliptical orbit

Arrival in moon orbit

GET 80:05 (10.37 p.m. BST 19 July)
SM retro-fire (LOI-2)

lunar north pole

62×75 miles

earth

Correction of orbit to near circle

weight 96,012 pounds [forty-three tons]. We're seven minutes and forty-five seconds away from the LOI [lunar orbit insertion]. With a good burn, the Madrid station should acquire Apollo Eleven at 76 hours: 15 minutes: 29 seconds [GET].

GET 76:04

PAO: 'We're about fifty seconds away from the acquisition time provided Apollo did not make the burn [and was on a course to bring it back to earth]. If Apollo Eleven achieved only a partial burn, we could receive a signal at any time.

GET 76:15:29

PAO: Madrid AOS [acquisition of signal].

MCC: *'Apollo Eleven, Apollo Eleven, this is Houston, can you read me?'*

PAO: Telemetry indicates that the crew is working on the antenna angles to bring the high gain antenna [attached to the service module] to bear.

MCC: *'Apollo Eleven, Apollo Eleven, do you read me?'* Apollo 11: *'Read you loud and clear, Houston, how us?'* MCC: *'Roger, reading you the same now. Could you repeat your burn status report. Send the whole thing again, please.'* Armstrong: *'It was like perfect . . . 44* [a computer programme] *showed us in a six zero, decimal nine by one six nine, decimal nine* [the parameters, in nautical miles, of their elliptical orbit].' All was well with Apollo, they had 'come round the corner' in the intended orbit. Soon they were describing the lunar surface from their vantage point above the moon. Armstrong: *'We're getting our first view of the landing approach. ... We're going over that Tarantius crater and the pictures and maps brought back by Apollos Eight and Ten give us a very good preview of what to look at. It looks very much like the pictures but, like the difference between watching a real football game and one on TV, there's no substitute for actually being here. ... We're going over the Messier series of craters right at this time, looking vertically down on them and, hey, we can see good-sized blocks in the bottom of the crater. I don't know what our altitude is now but those are pretty good-sized blocks.'* MCC: *'OK, just roughly*

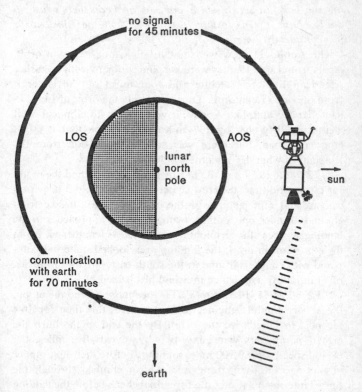

Problems of communication with the Apollo 11 spacecraft in lunar orbit: loss of signal (LOS) as the craft goes behind the moon on each orbit, acquisition of signal (AOS) as it reappears

it looks like you're one twenty miles or one thirty miles right now. . . . Make that one twenty-seven miles.'

GET 76:56 (7.28 p.m. BST) Aldrin: *'We just went into space-craft darkness. Now with earthshine the visibility is pretty fair. Looking back behind me now I can see the corona from where the sun has just set. When a star sets up here, there's just no doubt about it. One instant it's there and the next instant it's just completely gone.'*

The Apollo 11 crew were not totally alone in moon orbit. Apart from Luna 15, they were keeping company with an earlier American Explorer satellite and four burnt out Soviet craft from previous Luna shots. During the day the orbit of Luna 15 had changed and, like Apollo, it was now in an elliptical orbit measuring fifty-nine by 137 miles. Speculation that it might land continued, but there was still no indication from the Russians of what the outcome would be.

GET 78:26 (8.58 p.m. BST) Emerging from behind the moon for the second time, the crew switched on their colour television camera and sent pictures of the cratered lunar surface from about 100 miles up. As they swung over the approaches to the landing site near the terminator, the shadows lengthened. From their position in orbit, the landing area looked quite flat compared with rougher ground at the south-eastern edge of the Sea of Tranquillity over which they had just passed.

GET 80:04:51 (10.37 p.m. BST) Once more behind the moon, the service module engine was again fired, this time for 16·4 seconds, to circularize the orbit. By the end of the burn the orbital parameters were sixty-two by seventy-five miles, the orbital speed some 3,700 miles an hour. Before their first supper in moon orbit, Armstrong and Aldrin climbed through the tunnel into the lunar module for a final check before the landing attempt to be made the next day.

Handgrips mounted on the inside of the docking tunnel enabled Armstrong and Aldrin to haul their weightless bodies to the hatch leading into the ascent stage of the lunar module. The hatch was opened inwards by a single handle and the two astronauts floated head first into the cabin section of the craft. The cabin section or crew compartment was essentially a cylinder ninety-two inches in diameter and forty-two inches deep. Just above the floor at one end of this cylinder was the other hatch through which, the next day, Armstrong would pass into history. Immediately above this hatch were the flight controls and display panels and, to each side of these, triangular windows, one for the commander, Armstrong, on the left-hand side, the other for the lunar module pilot, Aldrin, on the right. Above Armstrong's post was an additional window, looking upwards and giving him a view of the command module as he separated the two craft or docked them together. All three lunar module windows were double paned. The outer pane was designed to withstand the impact of micrometeorite showers, while the inner one was coated with fifty-nine layers of metallic oxide to reduce infra-red and ultra-violet radiation. Each window was fitted with a heater to prevent fogging, as well as a roller shade and a glare shield.

Life in the lunar module was pretty cramped. Both sides of the crew section were filled with extra instrumentation and equipment bays. To hold the astronauts steady in zero-G, harnesses suspended from the top and sides of the crew section were provided. Behind them, on the floor of the cabin, was the top of the main rocket engine of the ascent stage. The fuel for this engine was housed behind the crew compartment in an unpressurized section which also contained the crew's oxygen and water. Beneath the control and display panels in front of Aldrin's station were stowed food, cameras, and other equipment including a multi-purpose tool which in dire emergency could have been used to take the cover off the ascent engine had they not been able to restart it on the lunar surface. Below Armstrong's station a similar equipment bay housed a spare environment control subsystem (ECS) for use had the main ECS atmosphere conditioner broken down. Behind the ascent engine cover further equipment bays contained the main ECS, a fire extinguisher, the cases in which the lunar samples would be placed, more food, the lunar boots which the astronauts would wear for the moon walk, and elements of the power and navigation equipment aboard the lunar module.

Also stowed in the pressurized cabin section were the two bulky backpacks – the portable life support systems (PLSS) which provided the astronauts' spacesuits with an earthly atmosphere and which cooled the water circulating through tiny tubes in the suits' innermost layers. An additional emergency supply of oxygen, enough for thirty minutes, was contained in two two-pound bottles. Also attached to the PLSS backpack were the communication aerials allowing Armstrong and Aldrin to speak to each other, to earth, and to Collins in the command module by relay through the lunar module's communication system.

The ascent stage was mounted on top of the octagonal descent stage, which consisted essentially of a rocket engine of variable thrust mounted on four legs. Weighing twice as much as the descent stage, it also housed the experimental packages to be left on the moon and the television camera which was to show Armstrong's first footprint on the moon.

Seeing it for the first time, many astronauts have been amazed by the flimsiness of the LM. Every unnecessary ounce was pared away by its manufacturers, and since it was not designed to fly in an earthly atmosphere, its skin is only a few thousandths of an inch thick. Yet in this tissue-paper spacecraft man was about to land on the moon.

GET 84:00 (2.32 a.m. BST, 20 July) Back in the command module with Collins, Armstrong and Aldrin were eating their last meal before the momentous day as they glided at 3,700 miles an hour above the hidden surface of the moon, out of sight of their home planet. Having sent back their final report for the day, they stretched themselves out for nine hours of sleep. Back on earth the ever watchful medical team at Mission Control reported them in perfect trim.

GET 93:28 (noon, BST) The sun was slowly rising over the Sea of Tranquillity after the thirteen-day-long lunar night as the Apollo 11 craft came swinging round on its tenth orbit of the moon. Armstrong, Aldrin and Collins were preparing their breakfast. *'We can pick out almost all of the features* [of the landing run-in] *we've identified previously,'* reported Armstrong. He and Aldrin had spent a restless night in anticipation of the day's events. Only Collins had slept the full eight hours. After breakfast Mission Control read them the news and then told them: *'An ancient legend says that a beautiful Chinese girl named Chango has been living there for four thousand years. It seems she was banished to the moon because she stole the pill of immortality from her husband. You might look for her companion, a large Chinese rabbit who is easy to spot since he's always standing on his hind feet in the shade of the cinnamon tree.'* Collins: *'We'll keep a close eye for that bunny girl.'*

GET 95:40 (2.12 p.m. BST) Armstrong dressed himself in his pressure suit (less helmet and gloves), opened the tunnel access hatch for Aldrin, as yet not in his spacesuit, who crawled into the lunar module cabin. At a word from Aldrin, Armstrong switched off the power line from the CM to the LM as Aldrin transferred the LM to its own internal power. While Armstrong disconnected the power line and stowed it in the CM, Aldrin switched the power into the lunar module's circuits, started up the mission timer on the display panel, and turned on the pump which circulated glycol through the cooling system within the module's electronic systems. Next he checked the caution and warning systems, the circuit breaker and the time base (the craft's electronic clock). As Armstrong came through the tunnel carrying his helmet and gloves with him, Aldrin was testing out the primary guidance navigation control system (PGNCS or 'pings'). The role of pings was to do as much of Armstrong's job for him as was possible. Backed by the onboard computer, it constantly measured and evaluated the flight of the lunar

module, correlated this data with the information it had about
its destination, and then, by automatically controlling either of
the craft's main engines and the sixteen attitude control thrusters
mounted in groups of four around the ascent stage, flew the
lunar module along the correct flight path. Armstrong was
able to override pings by flipping the guidance control switch
on his panel to 'aggs', the abort guidance system (AGS), which
gave him much more say in where, when and how fast he was
going.

Safely inside the lunar module, Armstrong had turned on
the lunar module's environment control subsystem and had
plugged his spacesuit into it, while Aldrin positioned the biomedical monitoring switch which would feed the medical data
originating from the small pads attached to their bodies from
the LM to the medical team at Mission Control. Aldrin then
switched on the LM's S-Band transmitter and checked that the
S-Band antenna was properly aligned with the earthbound
tracking station. He then went back through the tunnel into the
command module where he put on his pressure suit. In the
meantime Armstrong was checking VHF (radio) communication between the CM and the LM with Collins. Among other
checks he then carried out was the alignment of the LM's inertial measurement unit (IMU). The IMU was the heart of pings
and aggs, and in conjunction with the onboard computers it
showed Armstrong exactly where he was every fraction of a
second. Basically it consisted of three accelerometers, sensitive
to changes of speed in the three directions in which the LM
could go. Gyroscopes gave the IMU a fixed sense of position
and attitude in space with reference, for example, to that imaginary line between earth and moon. The onboard computer
remembered for how long the spacecraft had been travelling
and so between them the computers and the IMU knew exactly
where the craft was as well as where it would be at any given
moment. While the Apollo craft was behind the moon, only the
IMU could tell the astronauts that all was well.

As Armstrong finished aligning the IMU, Aldrin came back
into the LM carrying his helmet and gloves and plugged his
suit into the lunar module's ECS. Armstrong then checked that
the LM's drogue and the CM's probe were properly replaced in

the docking tunnel and secured the LM's docking hatch, while Collins checked that the hatch at the CM end of the tunnel was also secure. Aldrin checked the ascent stage battery and they both put on their helmets and gloves. Collins was also kitted out in his spacesuit with helmet and gloves and had let the air out of the docking tunnel. Armstrong and Aldrin then checked that the air supplies to their spacesuits and to the cabin were as they should be. This test completed, all three astronauts removed their helmets and gloves for the final phases of checkout before undocking the two craft.

Collins now manoeuvred the Apollo spacecraft to a position where he could track the LM with his radar once it had separated. Armstrong and Aldrin carried out further checks on the guidance system, the descent stage engine, the rendezvous radar and the LM's attitude control thrusters. As they completed their twelfth orbit of the moon, the Apollo 11 crew were ready to undock the two craft. By now two new call-signs were being heard in the conversations between the spacecraft and Mission Control. The lunar module had been codenamed Eagle after the American national bird, while the command module's call-sign was Columbia, in memory of the rocket which over 100 years earlier had propelled Jules Verne's legendary heroes to the moon from a site near Tampa in Florida only a little more than 100 miles from Cape Kennedy.

GET 99:22 (5.54 p.m. BST) MCC: '*Apollo Eleven, Houston, we're go for undocking, over.*' Eagle: '*Roger, understand.*' As the two craft, still linked, slipped behind the moon, Armstrong could be heard telling Collins: '*Hello, Columbia, systems looking good.*'

Armstrong and Aldrin were now strapped in their harnesses at their stations in the LM: Armstrong on the left, Aldrin on the right. All three crew members were now wearing helmets and gloves. Soon Collins would flip a switch releasing the last three of fifteen latches that held the two craft together, while back on earth Mission Control waited for them to reappear.

GET 100:14 (6.48 p.m. BST)

PAO: This is Apollo control at 100 hours, 14 minutes. We are now less than two minutes from reacquiring the spacecraft on the

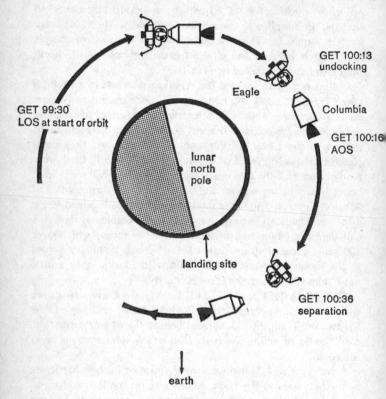

GET 100:13
undocking

Eagle

Columbia

GET 100:16
AOS

GET 99:30
LOS at start of orbit

lunar
north
pole

landing site

GET 100:36
separation

earth

Columbia and Eagle undock

thirteenth revolution. . . . We're presently twenty-five minutes away from the separation burn [using the attitude control thrusters of the service module] that will be performed by Mike Collins in the command module to give the LM and the CSM a separation of about two miles. . . . We'll stand by now to reacquire the spacecraft . . . We have acquisition of signal [telemetry].

MCC: *'Hello, Eagle, Houston. We're standing by, over.'* Eagle: *'Roger, Eagle, stand by.'* MCC: *'Roger, how does it look?'* Armstrong: *'The Eagle has wings.'* Safely undocked, Eagle and Columbia were flying side by side as they came from behind the moon.

   GET 100:32 (7.04 p.m. BST) MCC: *'Columbia, Houston, how do you read?'* Collins: *'I hear you loud and clear, Houston, how me?'* MCC: *'Roger, Mike. . . . On my mark, seven minutes to ignition. Mark, seven minutes.'* Collins: *'I got you . . . everything's looking real good.'*

PAO: This is Apollo control. We are six minutes, eight seconds from ignition.

MCC: *'Houston, you are looking good for separation. You are go for separation, Columbia, over.'* Collins: *'Columbia, understand. . . . We're really stabilized, Neil, I haven't fired a thruster in five minutes.'* Eagle: *'Mike, what's going to be your pitch angle at SEP* [separation]?*'* Collins: *'Zero zero seven degrees.'* Eagle: *'OK.'* Collins: *'I think you've got a fine-looking flying machine there, Eagle, despite the fact you're upside down.'* Eagle: *'Somebody's upside down.'* Collins: *'OK, Eagle, one minute until T* [separation burn time]. *You guys take care.'* Eagle: *'See you later.'*
   GET 100:40 (7.12 p.m. BST)

PAO: This is Apollo control. That separation was performed as scheduled. In the command module a Delta-V [speed change] of about two point five feet per second [was recorded], which gave a separation to the two vehicles of about eleven hundred feet at the beginning of the descent orbit insertion [DOI] manoeuvre [which was to take place behind the moon].

Eagle: *'Going right down US one, Mike.'* (US highway number one is the nickname the astronauts have given to the approach for the landing site on the Sea of Tranquillity).

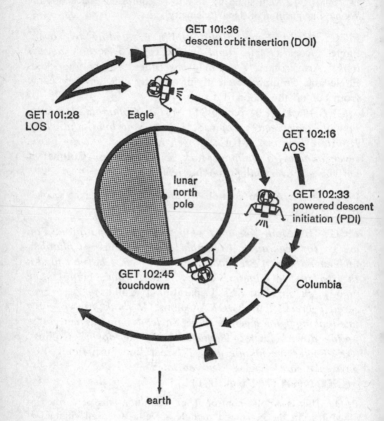

GET 101:36
descent orbit insertion (DOI)

GET 101:28
LOS

Eagle

lunar
north
pole

GET 102:16
AOS

GET 102:33
powered descent
initiation (PDI)

GET 102:45
touchdown

Columbia

earth

Descent to Tranquillity base

Armstrong and Aldrin were now carrying out last-minute checks before the descent orbit insertion burn (DOI) of the LM's descent engine which would put the LM into a sixty-six by ten mile orbit of the moon.

GET 101:18 (7.50 p.m. BST) MCC: *'Eagle, Houston, you are go for DOI, over.'* Eagle: *'Roger, go for DOI. Do you have AOS and LOS times?'* MCC: *'Roger, for you LOS at 101:28, AOS at 102:16, over.'* Eagle: *'Roger, copy.'*

GET 101:24 (7.56 p.m. BST) MCC: *'Eagle, on my mark we'll have twelve minutes to* [DOI] *ignition, over.'* Eagle: *'Roger.'* MCC: *'Eagle, Houston, stand by for my mark. Mark, twelve minutes to ignition, over.'* Eagle: *'We copy.'* MCC: *'Columbia, Eagle, Houston, three minutes to LOS. Both looking good going over the hill.'* Collins: *'Columbia, roger.'* Eagle: *'Eagle, rog.'*

GET 101:27 (7.59 p.m. BST)

PAO: This is Apollo control. We've had loss of signal now. When we next acquire the lunar module, it should be at an altitude of about eighteen nautical miles [twenty-one statute miles] on its way down to a low point of about 50,000 feet from where the powered descent [PDI] to the lunar surface will begin. As the spacecraft went around the corner, all systems on both vehicles looked very good. Everything is go here in Mission Control and aboard the spacecraft for the descent orbit insertion [DOI] to occur in six minutes, thirty-eight seconds. This is Apollo control, Houston, at 101 hours: 29 minutes.

Behind the moon, Armstrong was due to start up the descent engine of the lunar module and burn it for exactly 29.8 seconds. The 9,800-pound thrust of the engine was to slow the LM down and drop it into a lower orbit before the powered descent (PDI) to the landing site on the Sea of Tranquillity. Assembled at Mission Control were most of the senior NASA officials including Wernher von Braun, many of the astronauts and some 3,500 newsmen from all over the world. The wives of the crew had elected to stay at home in front of their television sets. Mission Control and much of the rest of the world, tuned in to the global television and radio link-up, held their breath as they waited for Eagle to reappear.

PAO: We're now coming up to thirty seconds to acquisition of the command module and we'll stand by for that event. ... Network controller says we have acquisition of [telemetry] signal from the command module.

MCC: '*Columbia, Houston, we're standing by, over, Columbia, Houston, over.*' Collins: '*Houston, Columbia, reading you loud and clear, hear me?*' MCC: '*Roger, Mike, how did it go? Over.*' Collins: '*Listen, babe, everything's going just swimmingly, beautiful.*' MCC: '*Great, we're standing by for Eagle.*' Collins: '*OK, he's coming around.*' MCC: '*We copy, out.*'

PAO: We have acquisition of [telemetry] signal from the LM.

Eagle: '*Houston, Eagle, how do you read?*' MCC: '*Eagle, we're standing by for your burn report, over.*' Eagle: '*Roger, the burn was on time.*' For a few minutes difficulties with communications occupied Mission Control as the big 210-foot dish at Goldstone in California lost 'lock' with the lunar module's S-Band antenna and some voice exchanges had to be relayed via Collins in the command module.

PAO: Guidance says we're go ... now twelve minutes, fifty-four seconds to ignition. ... Everything still looking good at this point. We presently show the LM at an altitude of 12·9 nautical miles [fifteen statute miles] and descending.

They were now approaching the low point of their new orbit, when they would fire the descent engine once more to bring them out of orbit to man's first landing on another body in the solar system.

PAO: Coming up on five minutes to ignition ...

MCC: '*Eagle, Houston, if you read, you're go for powered descent, over.*' Collins: '*Eagle, this is Columbia, they just gave you a go for powered descent.*' MCC: '*Columbia, Houston, we've lost them on the high gain [circuit] again. We recommend they yaw right ten degrees and reacquire.*' Collins: '*Eagle, this is Columbia, you're go for PDI and they recommend you yaw right and try the high gain again. Eagle, you read Columbia?*' Aldrin: '*Roger, read you.*' Collins: '*OK.*' MCC: '*Eagle, Houston, we read you now. You're go for PDI,*' over.' Aldrin: 'Roger, understand.' MCC: 'Eagle, Houston. On my mark, 3:30 to ignition. . . . Mark 3:30.' Aldrin: 'Roger, copy.' Aldrin then read off a checklist to Armstrong who was getting ready to fly the craft down to the Sea of Tranquillity.

PAO: Coming up to one minute to ignition.

GET 102:33 (9.05 p.m. BST) In front of Armstrong two greenish figures on the computer display panel said '63' (the computer landing programme). He had just five seconds to make up his mind whether to go for the landing or continue in orbit. Armstrong pressed the 'proceed' button. Immediately the descent engine burst into life at the beginning of a 300-mile-long arc down to the surface of the moon. The lunar module was flying face down, feet first about 50,000 feet above the western edge of the Sea of Fertility to the east of their landing site on the lunar equator in the south-western edge of the Sea of Tranquillity. For the first twenty-six seconds the descent engine burnt at only ten per cent of its maximum thrust to allow the delicately balanced craft to settle in the proper alignment for the landing flight path, then gradually the full power of the engine was turned on. Through the triangular windows Armstrong and Aldrin checked off the landmarks along 'US highway number one' as their height began to drop.

Beneath them the astronauts saw the crater Maskelyne as they swung over the Sea of Tranquillity. Aldrin: 'Aggs and pings agree very closely.'

PAO: Two minutes, twenty seconds [from PDI]. Everything looking good. We show an altitude of about 47,000 feet.

At 43,300 feet, on command from the pings computer, the lunar module rolled around to 'face-up', leaving the astronauts a view of the stars while the landing radar locked on to the ground. MCC: 'You're looking good at three, coming up . . . three minutes.' Height 39,500 feet; the radar was now feeding the instrument panels in front of the astronauts with their altitude second by second. MCC: 'Eagle, Houston, you are go. Take it all at four minutes. . . . You are go to continue powered descent.' Aldrin: 'Roger . . . the earth [is] right out

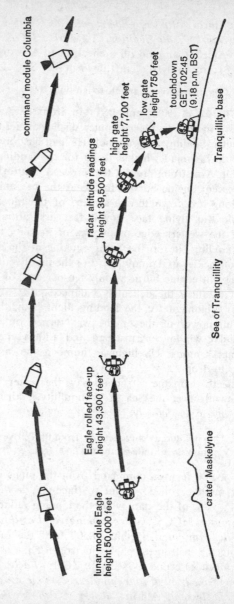

Landing at Tranquillity base, 20 July 1969

east

lunar module Eagle
height 50,000 feet

Eagle rolled face-up
height 43,300 feet

radar altitude readings
height 39,500 feet

command module Columbia

earth

high gate
height 7,700 feet

low gate
height 750 feet

touchdown
GET 102:45
(9.18 p.m. BST)

Tranquillity base

Sea of Tranquillity

crater Maskelyne

west

of our front window.' During the descent the alarm light on the LM control panel lit up repeatedly. This was later found to be due to overloading of the computer rather than serious trouble.

PAO: Good radar data. Altitude now 33,500 feet.

Gradually the LM was beginning to pitch up as it dropped nearer to the ground. Armstrong and Aldrin had now passed a point of no return, for there was no longer enough fuel in the descent engine to push the craft back up into orbit. Had they wanted to do this, they would have had to discard the descent stage and use the ascent stage engine to climb back up. At 24,639 feet the pings computer was programmed to cut the descent engine to sixty per cent thrust. *'Throttle down better than simulator,'* reported Aldrin.

PAO: Altitude now 21,000 feet. Still looking very good. Velocity down now to 1,200 feet per second [816 miles an hour].

The radar was now feeding 'rate of descent' information to Armstrong and Aldrin; the LM was pitched up far enough for them to be able to see the lunar surface again.

PAO: Seven minutes, thirty seconds into the burn. Altitude 16,300 feet.

The lunar module was now only about ten miles from the chosen landing site. MCC: *'Stand by. . . . You're looking great at eight minutes.'*

PAO: Altitude 9,200 feet.

Pings was flying the LM to a point in space 7,700 feet up and five miles to the east of the landing site, called 'high gate', at which the landing site first came into view as they sped along towards it at some 370 miles an hour, losing height at some 132 feet per second. Armstrong now began to take stock of his position.

MCC: *'Eagle, you're looking great, coming up . . . nine minutes.'*

PAO: We're now in the approach phase ... looking good. Altitude 5,200 feet.

Aldrin: '*Manual auto attitude control is good.*' Armstrong was controlling the sixteen attitude control thrusters mounted around the ascent stage of the lunar module, now only pitched over some fifty degrees from vertical. MCC: '*Houston, you're go for landing, over.*' Aldrin: '*Roger, understand, go for landing … 3,000 feet … we're go, hang tight … we're go … 2,000 feet … into the aggs* [Armstrong had now virtually taken over from the automatic landing sequence] *… 47 degrees* [pitch from vertical].' MCC: '*Roger, you're looking great, you're go.*'

PAO: Altitude: 1,600 feet … 1,400 feet … still looking very good.

The LM was now approaching what should have been 'low gate', a point some 500 feet above and 2,000 feet east of the landing site. Pitched to nearly the vertical position, the LM was now 'standing' on its descent engine, dropping gently towards the landing area. Through his triangular window Armstrong could now see where aggs was taking him: straight into a crater full of large boulders and 'big enough to house the Houston Astrodome'. They were four miles beyond the intended position, due to a slightly late PDI rather than to any interference from 'mascons', concentrations of mass hidden below the lunar surface. They are thought to be huge meteorites, of density greater than the rest of the lunar rock, which have been buried after impact, and give rise to a sudden increase in gravity that can draw spacecraft in low orbits perilously off course. Until the gravity maps of the moon are more perfect, even computers cannot guard against this peril.

Armstrong flicked a switch to give him full manual control and, summoning up all his flying experience, guided the ungainly monster to safety. With his right hand he controlled the sixteen attitude control thrusters while his left hand adjusted the thrust of the descent engine. Armstrong remained ice cool, his heart rate a steady determined pace, as he edged the LM into the last few seconds towards touchdown, checking its progress on the attitude indicator, the altimeter, the various speed gauges, and with quick glances out of the window. Aldrin: '*35 degrees* [pitch] *… 35 degrees … 750* [altitude] *… coming down at 23* (feet per second) *… 700 feet, 21 down*

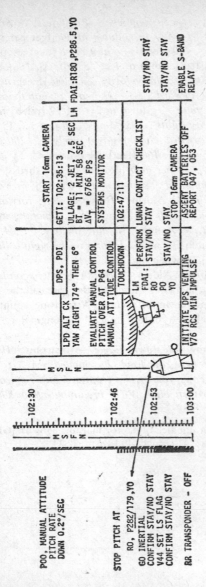

Page from the Apollo 11 flight plan for the moment of touchdown. The left-hand column refers to Collins in the command module. The middle column refers to Armstrong, the right-hand column to Aldrin, both in the lunar module.

[feet per second] . . . *33 degrees . . . 600 feet, down at 19*
[feet per second] . . . *540 feet, down at 30* [feet per second]
. . . *down at 15 . . . 400 feet, down at 9 . . . 8* [degrees, pitched]
forward . . . 350 [feet], *down at 4 . . . 330* [feet], *3½ down . . .*
we're pegged on horizontal velocity . . . 300 [feet], *down 3½*
. . . 47 [degrees] *forward . . . 1½ down . . . got the shadow*
[of the LM] *out there . . . down at 2½ . . . 19* [pitch] *forward*
. . . altitude velocity lights . . . 3½ down, 220 feet . . . 13 [pitch]
forward . . . 11 forward, coming down nicely . . . 200 feet, 4½
down . . . 5½ down . . . 160 [feet], *6½ down . . . 5½ down, 9*
[pitch] *forward . . . 5 per cent* [descent engine thrust] *. . . 75*
feet, things looking good . . . down a half, 6 [pitch] *forward . . .'*

MCC: '*60 seconds.*' Aldrin: '*Down 2½ . . . forward . . .*
forward . . . good . . . 40 feet, down 2½ . . . picking up some
dust . . . 30 feet, 2½ down . . . faint shadow . . . 4 [pitch]
forward . . . 4 [pitch] *forward, drifting to the right a little . . .*
6 [forward pitch], *down a half.*'

MCC: '*30 seconds.*' Aldrin: '*Drifting right . . . contact light*
[landing probes attached to three of the pads of the descent
stage had touched the lunar surface, Armstrong counted one
second and punched a button to cut the engine] *. . . OK,*
engine stop . . .'

MCC: '*We copy you down, Eagle.*' Armstrong: '*Houston,*
Tranquillity base here, the Eagle has landed.' MCC: '*Roger,*
Tranquillity, we copy you on the ground. You've got a bunch
of guys about to turn blue. We're breathing again. Thanks a
lot.'

PAO: We have an unofficial time for that touchdown: [GET] 102
hours: 45 minutes: 42 seconds [9.18 p.m. BST].

Avoiding the boulder-filled crater he had sighted, Armstrong had landed Eagle some four miles down range from the originally selected landing site. It was some time before Mission Control could figure out exactly where they were. Armstrong's cool had given way to excitement as he put Eagle down on the rocky lunar surface; his heart rate had risen to 156 beats per minute, but now it was back down to 99 as he and Aldrin went through a rapid check of the spacecraft and a simulated countdown in readiness for an emergency take-off should this be necessary. Even as Armstrong and Aldrin were preparing to land, Russia's Luna 15 was adding to the drama. It too had swung into a low orbit, skimming ten miles or so above the moon on its nearest approach. It is almost certain that it was intended to make a soft landing, but the mystery came to an abrupt end when Sir Bernard Lovell reported from Jodrell Bank that the craft had plunged into the Sea of Crises (rather fittingly) at some 300 miles an hour, and later Tass added that its mission 'had been completed'.

For a few minutes, while Collins was still overhead in the command module, Armstrong could choose to fire the ascent engine and rejoin him in orbit. In the excitement of the landing, Collins had almost been forgotten, and it was in an almost plaintive voice that he called up Mission Control:

'*Houston, do you read Columbia on the high gain?*'

MCC: '*Roger, we read you, Columbia. He has landed Tranquillity base. Eagle is at Tranquillity, over.*'

Collins: '*Yeah, I heard the whole thing . . . fantastic.*'

Eagle had now been down for seven minutes and the time

was approaching for the last stay or no-stay decision before they would commit themselves to remaining on the surface while Collins made at least one orbit of the moon.

MCC: '*Eagle, Houston, you are stay for T2, over.*'

Armstrong: '*Have your stay for T2, we thank you.*' This gave Armstrong another few minutes to change his mind. He now found time to report on his final approach to the landing site: '*Houston, that may have seemed like a very long final phase. The auto-targeting was taking us right into a football-field-sized crater, with a large number of big boulders and rocks for about one or two craters' diameters around us, and it required flying manually over the rock field to find a reasonably good area.*'

MCC: '*Roger, we copy, it was beautiful from here, Tranquillity, over.*'

Armstrong: '*We'll get the details of what's around here, but it looks like a collection of just about every variety of shapes, angularities, granularities, every variety of rock you could find. The colours vary pretty much depending on how you're looking.* [Sunlight on the lunar surface is very stark and everything appears in sharp contrast, making it difficult to see any tone ranges or colour.] *There doesn't appear to be too much of a general colour at all, however it looks as though some of the rocks are boulders, of which there are quite a few in the near area, it looks as though they're going to have some interesting colours to them, over.*'

MCC: '*Roger, copy. Sounds good to us, Tranquillity. We'll let you press on through the simulated countdown, and we'll talk to you later, over.*'

Armstrong: '*OK. This one-sixth G is just like the airplane.*' (On training flights in planes flying a certain parabolic path it was possible to simulate weightlessness and varying degrees of gravity for short periods.)

MCC: '*Be advised that there's lots of smiling faces in this room, and all over the world.*'

Armstrong: '*There's two of them up here.*'

Collins: '*And don't forget one in the command module ... and thanks for putting me on* [communication] *relay, Houston, I was missing all the action.*'

Armstrong: *'Just keep that orbiting base ready for us up there now.'*

Collins: *'Will do.'*

PAO: We've just got a report that the LM systems looked good after landing. We're about twenty-six minutes now from loss of signal from the command module.

Armstrong and Aldrin, now committed to staying for at least two hours, had removed their helmets and gloves.

Armstrong: '[Out of] *the window is a relative plain cratered with a fairly large number of craters of five- to fifty-foot radius and* [there are] *ridges twenty, thirty feet high I would guess, and literally thousands of little one- and two-foot craters around the area. We see some angular blocks out several hundred feet in front of us that are probably two feet in size and have angular edges. There is a hill in view, just about on the ground track ahead of us, difficult to estimate* [the distance] *but might be half a mile or a mile. . . . It was really rough over the targeted landing area . . . extremely rough, cratered and large numbers of rocks that were probably larger than five or ten feet in size. . . . I'd say the colour of the local surface is very comparable to what we observed from orbit at this sun angle* [it was sunrise]. *It's pretty much without colour. It's grey and it's a very white chalk-grey as you look into the zero phase line* [directly opposite the sun], *and it's considerably darker grey, more like ashen grey as you look up ninety degrees to the sun. Some of the surface rocks in close here that have been fractured or disturbed by the rocket engine are coated with this light grey on the outside, but when they've been broken they display a dark, very dark grey interior and it looks like it could be country basalt* [an earthly volcanic rock].'

PAO: This is Apollo control. We've had loss of signal now from the command module.

At Tranquillity base Armstrong and Aldrin continued their work on the lunar module, preparing to power down the descent stage and readying the ascent stage for flight.

PAO: This is Apollo control at 103 hours, 44 minutes [10.16 p.m. BST, 20 July]. We have some updated information on the landing point. It appears that the spacecraft Eagle touched down just about on the lunar equator at 23.46 degrees longitude, which would put it about four miles from the targeted landing point down range. At this point all LM systems seem to look very good.

Tranquillity base: '*Houston, Tranquillity base is ready to go through the power down and terminate the simulated count-down.*'

MCC: '*Roger, stand by. ... Hello, Tranquillity base, you can start your power down now, over.*'

Tranquillity base: '*Roger, it has been started.*'

MCC: '*At Tranquillity base the white team* [referring to the 'white team' at Cape Kennedy which launches the Saturns] *is going off now and let the maroon team take over. We appreciate the great show. It was a beautiful job, you guys.*'

Tranquillity base: '*A recommendation at this point,* [we are] *planning an EVA* [extravehicular activity on the lunar surface] *with your concurrence starting about three hours from now.*'

MCC: '*We will support it* [your plan]. *We are go at this time. You guys are getting prime time TV there.*'

Tranquillity base: '*Hope that little TV set works.*' A small television camera attached to a panel in the descent stage of the lunar module was to relay the first step on the moon to an audience estimated to be 600 million, about one-fifth of the world's population.

GET 105:00 (11.32 p.m. BST) Armstrong and Aldrin were eating man's first meal on the moon prior to starting their preparations for the moon walk. In the original plan for the Apollo 11 flight, this meal was to have been followed by a four-hour rest period. How anyone had imagined that the astronauts would be able to sleep with the most exciting moment of their lives just the other side of a spacecraft hatch remains something of a mystery.

Their meal over, Armstrong and Aldrin were ready to put on their moon suits. The basic white spacesuit which they were wearing as they came in to land consisted of fifteen layers of special fabric, individually tailored for each astronaut. The outermost layer of the $100,000 garment was made of 'Super

116

Beta'. This was woven from fibreglass, each filament of which was coated with a tough, nylon-type plastic called Teflon. The Super Beta fabric was then covered with two layers of strong plastic film. Called the integrated thermal meteoroid garment, the extravehicular spacesuit kept the astronaut warm and protected him from possible showers of meteorites that can be expected on the moon. It was well equipped with sockets into which were plugged the tubes from the PLSS supplying oxygen for breathing, water to keep him cool as it circulated through a network of tiny tubes next to his skin, and the communication systems through which he could speak to his companion on the lunar surface, as well as to Mission Control and the command module by way of the LM. When he was wearing the complete suit in the command module or the LM, these umbilical tubes were connected to the spacecraft's environment control system and communications equipment.

The gloves, tailored to fit plaster casts of the astronaut's hands, were locked into metal cuffs. The basic helmet fitted to a similar metal ring around the neck. To allow a degree of mobility while still affording maximum protection from the harsh lunar environment, the suit was jointed at the ankles, hips, elbows and shoulders. To complete the resemblance to a medieval knight fully kitted in armour plus a frogman's outfit, the moon walkers wore, in addition to this basic suit, a second helmet with two visors, a pair of over-gloves, and a pair of lunar boots fitted over the integrated boots of the pressure suit. The complete moon suit and backpack weighed more than the astronaut, but under one-sixth gravity the load was considerably lightened. While Armstrong and Aldrin were preparing to climb out of the LM, Collins had reappeared from behind the moon and was trying to catch a sight of them, but since it was still uncertain exactly where he should look, he passed over the landing area without seeing anything.

Their preparations almost complete, Armstrong and Aldrin tested the radio circuits between their suits and the lunar module, and switched on the television camera to see if it had survived the landing. As yet there was nothing for it to photograph, but it was working and Mission Control was receiving its signal.

117

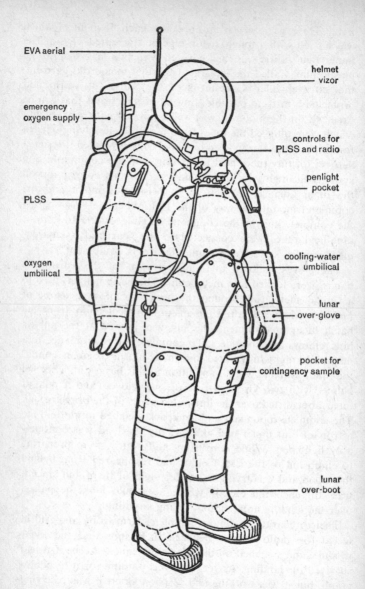

EVA aerial

helmet
vizor

emergency
oxygen supply

controls for
PLSS and radio

penlight
pocket

PLSS

oxygen
umbilical

cooling-water
umbilical

lunar
over-glove

pocket for
contingency sample

lunar
over-boot

Apollo 11 lunar spacesuit

Armstrong: *'Houston, this is Tranquillity, we're standing by for a go for cabin depress, over.'*

MCC: *'Tranquillity base, this is Houston, you are go for cabin depressurize, over.'*

GET 109:04 (3.36 a.m. BST, 21 July) Armstrong: *'Everything is go here. We're just waiting for the cabin pressure to bleed, to blow enough pressure to open the hatch.'*

MCC: *'Roger, we're showing a real low static pressure on your cabin. Do you think you can open the hatch at this pressure?'*

Armstrong: *'We're going to try it . . . hatch coming open.'*
Aldrin now guided Armstrong as he made his ungainly exit, on his knees and backwards through the narrow hatch, into an alien world.

Aldrin: *'Neil, you're lined up nicely. . . . Toward me a little bit. . . . OK, down . . . roll to the left . . . put your left foot to the right a little bit . . . you're doing fine.'*

Armstrong: *'OK, Houston, I'm on the porch.'* Unlike all the moon landing heroes of science fiction, Armstrong was crouching with his back to the moonscape on a platform just outside the hatch. Now he gingerly felt with his feet for the first rungs of the ladder.

PAO: Neil Armstrong on the porch at 109 hours, 19 minutes, 16 seconds [3.51 BST] . . . twenty-five minutes of PLSS time expended now.

The life support systems Armstrong and Aldrin were now living on had enough oxygen and water for up to four hours.

Aldrin: *'OK, everything's nice and sunny in here.'*

Armstrong: *'OK, can you pull the door open a little more?'*

Aldrin: *'Did you get the MESA out?'* The MESA (modularized equipment stowage assembly) was a panel on the descent stage to the left of the foot of the ladder which Armstrong opened by pulling a cord. As the flap came down, it revealed the television camera and some of the equipment which would be used by the astronauts to collect moon samples.

Armstrong: *'I'm going to pull it now. Houston, the MESA came down all right.'*

MCC: *'Houston, roger, we copy and we're standing by for*

119

your TV.... Man, we're getting a picture on the TV.' On the
giant screen at Mission Control an obscure jumble of black
and white images flickered into place. MCC: '*There's a great
deal of contrast in it and currently it's upside down on our
monitor, but we can make out a fair amount of detail.... OK
Neil, we can see you coming down the ladder now.*'

On the television screen the picture had now turned the
right way up. The surface of the moon appeared as a sheet of
white paper while the 'sky' was inky black. In the foreground
the ladder could be made out, a grey form descending from
top left of the screen. A shadowy grey figure could be seen coming down it.

Armstrong: '*I'm at the foot of the ladder. The LM foot
pads are only depressed in the surface about one or two inches.
Although the surface appears to be very fine, fine grained, as
you get close to it, it's almost like a powder. Now and then
it's very fine.*' Armstrong was now standing with both feet on
the three-foot diameter pad at the base of the landing leg.

Armstrong: '*I'm going to step off the LM now.*' He lifted
his left foot and, pushing it out a little, planted history's most
witnessed footprint. '*That's one small step for a man, one giant
leap for mankind.*'

Armstrong: '*The surface is fine and powdery. I can pick it
up loosely with my toe. It adheres in fine layers like powdered
charcoal to the sole and sides of my boots. I only go in a small
fraction of an inch, maybe one-eighth of an inch, but I can
see the footprints of my boots and the treads in the fine, sandy
particles. There seems to be no difficulty in moving around, as
we suspected. It's even perhaps easier than the simulations at
one-sixth G that we performed in the simulations on the
ground. It's actually no trouble to walk around. The descent
engine did not leave a crater of any size. There's about one
foot clearance on the ground* [from the bottom of the descent
engine nozzle]. *We're essentially on a very level place here. I
can see some evidence of rays emanating from the descent
engine, but* [a] *very insignificant amount. OK, Buzz, we're
ready to bring down the camera.*' Aldrin now lowered the still
camera on a pulley attached to a rope.

Armstrong: '*It's quite dark here in the shadow* [of the LM]
120

Page from the Apollo 11 flight plan covering the moment when Armstrong opened the lunar module hatch and descended to the moon's surface. Armstrong stepped on the moon earlier than originally planned.

DON GLOVES

PRESSURE INTEGRITY CHECK
PLSS O₂ ON SET CHRONOMETER
START EVA 0+00

FINAL PRE-EVA OPERATIONS
DEPRESS CABIN
FINAL SYSTEMS CHECKS   PLSS H₂O ON
OPEN FWD HATCH

0+10

INITIAL EVA
EGRESS TO PLATFORM
RELEASE MESA
DESCEND LADDER
REST/CHECK EMU SYSTEM
ASSIST AND MONITOR CDR
TURN TV ON
ACT 16mm CAMERA

UPDATE CSM
LM ACQUISITION
TIME

0+20

ENVIRONMENTAL FAMILIARI-
ZATION
CHECK STABIL, MOBIL, EMU

CONT SAMPLE COLLECTION
COLLECT AND STOW SAMPLE
MONITOR CDR
OPERATE 16mm CAMERA

0+30

M
S
F
N

112:30

112:39

112:45

113:00

M
S
F
N

CREW STATUS REPORT (SLEEP)
SELECT COMM NORMAL
LUNAR CONFIGURATION

COPY PAD
(SEE GET 114:20)

and a little hard for me to see if I have a good footing. I'll work my way over into the sunlight here without looking directly into the sun.' Armstrong had now stepped a few feet away from the LM opposite the ladder.

PAO: Time of the first step: 109:24:20 [3.56 a.m. BST].

Armstrong moved back into the shadow of the LM. *'Looking up at the LM, I'm standing directly in the shadow now looking up at Buzz in the window* [Aldrin was operating the sixteen-millimetre camera]. *I can see everything quite clearly. The light is sufficiently bright, backlighted into the front of the LM, that everything is very clearly visible.'*

PAO: The [flight] surgeon says that the crew is doing well. Data is good.

Armstrong had now secured the camera on a bracket mounted on his chest. *'I'll step out and take some of my first pictures here.'*
MCC: *'Roger, Neil, we're reading you loud and clear. We see you getting some pictures and the contingency sample. . . . Neil, this is Houston, did you copy about the contingency sample? Over.'*

Armstrong: *'Rog, I'm going to get that just as soon as I finish these pictures.'* Mission Control were obviously more concerned about the sample. If for any reason Armstrong had had to climb back into the LM in a hurry, he might not have had time to collect a sample of moon rock.

MCC: *'OK, going to get the contingency sample now, Neil?'*
Armstrong: *'Right.'* He reached down to a pocket just below the knee of his left leg and fished out a telescoped rod with a scoop attached to one end. He pulled out the rod and began to scrape up the first sample of the lunar surface. *'This is very interesting. It's a very soft surface but here and there where I plug with the contingency sample collector, I run into very hard surface, but it appears to be very cohesive material of the same sort. I'll try to get a rock in here. Here's a couple.'*

Aldrin: *'That looks beautiful from here, Neil.'*

Armstrong: '*It has a stark beauty all its own. It's like much of the high desert of the United States. It's different but it's very pretty out here. Be advised that a lot of the rock samples out here, the hard rock samples, have what appear to be vesicles in the surface* [bubbles in the rock which on earth would be associated with gases bubbling out of a hot lava to form rocks like pumice stone]. *Also I am looking at one now that appears to have some sort of phenocryst* [a mineral found in volcanic rocks on earth]. *I'm sure I could push it* [the sampling rod] *in further, but it's hard for me to bend down further than that.*' Aldrin, who was watching him, now helped him to get the sample into the pocket below his left knee. First Armstrong detached the scoop part of the rod and threw away the handle. Armstrong: '*You can really throw things a long way out there. That pocket open, Buzz?*'

Aldrin: '*Yes, it is, but it's not up against your suit. Hit it back once more. More toward the inside. OK, that's good.*'

Armstrong: '*That in the pocket?*'

Aldrin: '*Yes, push down, got it? No, it's not all the way in. Push it. There you go.*' Thus was the first piece of extra-terrestrial rock collected by man. Back on earth hundreds of scientists breathed a sigh of relief. For them this was the real object of the Apollo 11 flight.

Aldrin: '*I have got the* [sixteen-millimetre film] *cameras on at one frame a second.*'

Armstrong: '*Are you getting a TV picture now, Houston?*'

MCC: '*Neil, yes, we're getting a TV picture here. It's the first time we can see the bag on the LEC* [lunar equipment conveyor, the rope and pulley down which Aldrin was passing equipment to Armstrong] *being moved by Buzz. Here you come into our field of view.*'

Aldrin: '*Are you ready for me to come out?*'

Armstrong: '*Yes, just stand by a second. First let me move that* [the LEC] *over the edge for you.*'

Aldrin: '*That's got it. Are you ready?*'

Armstrong: '*All set. You saw what difficulties I was having. I'll try to watch your PLSS from underneath here. . . . OK, looks like it* [the PLSS] *is clearing OK. The shoes are about to come over the sill, OK, now drop your PLSS down. There you go, you're clear and spidery, you're good. About an inch clearance on top of your PLSS.'*

Aldrin: *'OK, you need a little bit of arching of the back to come down. How far are my feet from the. . . .'*

Armstrong: *'OK, you're right at the edge of the porch.'*

PAO: Forty-five minutes PLSS time expended.

MCC: *'Neil, this Houston, based on your camera transfer with the LEC, do you see any difficulty in SRC transfer?'* (SRC, sample return containers or lunar suitcases in which the samples would ride back to the lunar receiving laboratory in Houston. They had to be winched back up to the LM after they had been filled.)

Armstrong: *'Negative.'*

Aldrin: *'Now I want to back up partially and close the hatch, making sure not to lock it on my way out.'*

Armstrong: *'That's a good thought.'*

Aldrin: *'That's our home for the next couple of hours and I want to take good care of it. OK, I'm on the top step and I can look down over the RCU [remote control unit] and the landing gear pads. That's a very simple matter to hop down from one step to the next.'*

Armstrong: *'Yes, I found it to be very comfortable and walking is also very comfortable. You've got three more steps and then a long one.'*

Aldrin: *'OK, I'm going to leave that one foot up there and both hands down to about the fourth rung up.'*

Armstrong: *'There you go.'*

Aldrin: *'OK, now I think I'll do the same.'*

Armstrong: *'A little more, about another inch. There, you got it. That's a good step, about a three-footer.'* The two astronauts were now together on the surface.

Aldrin: *'Beautiful, beautiful.'*

Armstrong: *'Isn't that something. Magnificent sight down here.'*

Aldrin: *'Magnificent desolation. Looks like the secondary strut [of the landing gear] has little thermal effects on it right here, Neil.'*

Armstrong: *'Yeah, I noticed that. Isn't it fun.'*

Aldrin now tried moving at one-sixth gravity, executing little leaps and bouncing about.

Armstrong: *'You're standing on a rock, a big rock there now.'*

Aldrin: *'I wonder if that* [dent in the surface] *right under the engine is where the probe might have hit?'* (He was referring to a five-foot stick-like probe attached to the landing pad which signalled Armstrong that he had made contact with the lunar surface as he was landing.)

Armstrong: *'Yes, I think that's a good representation of our sideward velocity at touchdown there.'*

Aldrin: *'Can't say too much for the visibility here without the visor up.* [The visor, coated with a metallic oxide, was designed to cut down glare but made it difficult to see detail in shadow.] *The rocks are rather slippery . . . slide over it rather easily. Got to be careful you are leaning over in the direction you want to go. Neil, didn't I say we might see some purple rocks?'*

Armstrong was now about to move the TV camera. *'Find the purple rocks?'*

Aldrin: *'Yes, they are pretty small sparkly fragments. At a first guess some sort of biotite* [another mineral associated with igneous and metamorphic rocks]. *We'll leave that to the lunar analysis.'*

Armstrong: *'OK, Houston, I'm going to change lenses on you.'*

MCC: *'Roger, Neil.'*

As Armstrong changed the lens, the television screen on earth went briefly through a surrealistic lunar light show and then the picture settled, showing a narrower field of view but still the same subject, the ladder.

Armstrong: *'OK, Houston, tell me if you're getting a new picture.'*

MCC: *'Neil, this is Houston, that's affirmative. We're getting a new picture. You can tell it's a longer focal length lens and, for your information, all LM systems are go.'*

Armstrong: *'We appreciate that, thank you.'*

Aldrin: *'Neil is now unveiling the plaque.'* The television picture showed both astronauts in close-up standing in front of the ladder. Armstrong had peeled away some silver plastic to reveal the plaque.

HERE MEN FROM THE PLANET EARTH
FIRST SET FOOT UPON THE MOON
JULY 1969, A. D.
WE CAME IN PEACE FOR ALL MANKIND

NEIL A ARMSTRONG
ASTRONAUT

MICHAEL COLLINS
ASTRONAUT

EDWIN E. ALDRIN, JR.
ASTRONAUT

RICHARD NIXON
PRESIDENT, UNITED STATES OF AMERICA

Armstrong: *'For those who haven't read the plaque, we'll read the plaque that's on the front landing gear of this LM. First there's two hemispheres, one showing each of the two hemispheres of the earth. Underneath it says: "Here man from the planet earth first set foot upon the moon, July 1969 AD. We came in peace for all mankind." It has the crew members' signatures and the signature of the President of the United States.'* The two astronauts then moved the television camera to a position north of the lunar module and set it up to give a panoramic view of the rest of the activities on the lunar surface.

Aldrin: *'Houston, how close are you able to get things in focus?'*

MCC: *'This is Houston. We can see Buzz's right hand. It is somewhat out of focus. I'd say we're approaching down* to probably about eight inches to a foot behind the position where he is pulling out the cable [for the television camera].'

Aldrin: 'Forty . . . fifty feet. . . . Why don't you turn around and let them get a view from there and see what the field of view looks like? Turn around, to your right, I think, would be better.'

Armstrong: 'I don't want to get into the sun if I can avoid it.'

Aldrin: 'Houston, how's that field of view going to be?' The camera was now showing the whole LM, its long shadow falling across the surface to the west, looking like some gigantic ornate pepper pot. The quality of the picture was not very good and it was not possible to make out the finer detail of the LM, but its size, towering some twenty feet above the astronauts' heads, was very impressive.

MCC: 'Neil, this is Houston. The field of view is OK. We'd like you to aim it a little more to the right, over.'

Armstrong: 'Do you think I ought to be farther away or closer?' He swung the camera to show the lunar horizon around the landing site.

MCC: 'We'll line you up again when you finish the panorama.' For a moment the picture became blurred. 'Now you're going too fast on the panorama sweep.' Armstrong then steadied the camera to show a stony-looking plain stretching off into the north towards the centre of the Sea of Tranquillity.

MCC: 'We've got a beautiful picture, Neil.'

Armstrong: 'OK, I'm going to move it.'

MCC: 'OK, here's another good one. OK, we got that one.'

Armstrong: 'OK, now this one is right down front straight west and I want to know if you can see an angular rock in the foreground.'

MCC: 'Roger, we have a large angular rock in the foreground and it looks like a much smaller rock a couple of inches to the left of it. Over.'

Armstrong: 'All right, and then on beyond it about ten feet is an even larger rock that's very rounded. That rock, the closest one to you, is sticking out of the sand about one foot. Armstrong now swung the camera to look directly south across the shadow of the LM. Another desolate plain came into view.

Armstrong: *'The little hill just beyond the shadow of the LM is a pair of elongated craters. The pair together* [measures] *forty feet long and twenty feet across, and they're probably six feet deep. We'll probably get some work in there later.'*

MCC: *'Roger, we see Buzz going about his work.'* Aldrin was setting up the solar wind experiment, marking man's first attempt to use the moon as a solar observatory. This consisted of a panel of very thin aluminium foil which he took out of a tube and unrolled. The foil, attached to a rod which he pushed into the ground, was designed to trap particles emanating from the sun such as atoms of the gases helium, argon, neon, krypton and xenon, the so-called 'noble' elements. While Aldrin was setting this up, Armstrong fixed the television camera in its final position showing the LM in the top left of the screen and the astronauts at work around it.

PAO: One hour, seven minutes PLSS time expended.

Aldrin: *'Some of these small depressions* [are about] *three inches* [in depth].'

Armstrong: *'I noticed in the soft spot where we had foot-prints nearly an inch deep that the soil is very cohesive and it will retain a slope of probably seventy degrees.'*

PAO: Neil Armstrong has been on the lunar surface now almost forty-five minutes.

The command module now reappeared from behind the moon and was coming up over the landing site.

Collins: *'Houston, AOS.'*

MCC: *'Columbia, this is Houston reading you loud and clear, over.'*

Collins: *'Yes. This is history. Read you loud and clear, how's it going?'*

MCC: *'The EVA is progressing beautifully. I believe they are setting up the flag now.'*

Collins: *'Great.'*

MCC: *'I guess you're the only person around that doesn't have TV coverage of the scene.'*

Collins: 'That's all right, I don't mind a bit. How's the quality of the TV?'

MCC: 'Oh, it's beautiful, Mike, really is.'

Collins: 'Oh gee, that's great. Is the lighting halfway decent?'

MCC: 'Yes indeed, they've got the flag up now and you can see the Stars and Stripes on the lunar surface.'

Collins: 'Beautiful, just beautiful.'

The Stars and Stripes were now unfurled, thanks in part to Congress which had obstructed a NASA plan to set up the flag of the United Nations. For a brief moment the two men stood silently by their flag, as others had done on the summits of the highest mountains, by the sources of great rivers and at the frozen poles. Aldrin then set off on a series of hops, skips and jumps to try to find out the best method of locomotion on the moon.

Aldrin: 'OK, you do have to be rather careful to keep track of where your centre of mass is. Sometimes it takes about two or three paces to make sure you've got your feet underneath you. About two to three or maybe four easy paces can bring you to a nearly smooth stop. Next [to change] direction, like a football player, you just have to split out to the side and cut back a little bit. One [way of walking] called a kangaroo hop does work but it seems that your forward mobility is not quite as good.'

Armstrong: '[The] kangaroo hop does work, but it seems that your forward ability is not quite as good as it is in the conventional one foot after another [way of walking].'

MCC: 'Tranquillity base, this is Houston. Could we get both of you on the camera for a minute, please.'

Armstrong: 'Say again, Houston.'

MCC: 'Roger. We'd like to get both of you in the field of view of the camera for a minute. Neil and Buzz, the President of the United States is in his office now and would like to say a few words to you, over.'

Armstrong: 'That would be an honour.'

MCC: 'Go ahead, Mr President, this is Houston, out.'

President Nixon: 'Neil and Buzz, I am talking to you by telephone from the Oval Room at the White House and this

129

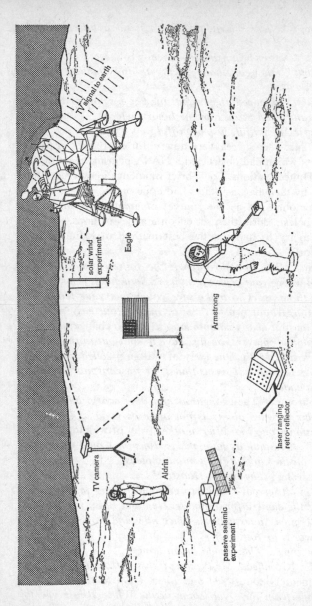

Armstrong and Aldrin on the lunar surface

certainly has to be the most historic telephone call ever made. I just can't tell you how proud we all are of what you. . . . For every American, this has to be the proudest moment of our lives, and for people all over the world, I am sure they too join with Americans in recognizing what a feat this is. Because of what you have done, the heavens have become part of man's world, and as you talk to us from the Sea of Tranquillity, it inspires us to double our efforts to bring peace and tranquillity to earth. For one priceless moment, in the whole history of man, all the people on this earth are truly one. One in their pride in what you have done, and one in our prayers that you will return safely to earth.'

Armstrong: *'Thank you, Mr President. It's a great honour and privilege for us to be here representing not only the United States but men of peace of all nations and with interest and a curiosity and a vision for the future. It's an honour for us to be able to participate here today.'*

President Nixon: *'And thank you very much and I look forward, all of us look forward to seeing you on the* Hornet [the recovery ship in the Pacific] *on Thursday.'*

Armstrong: *'Thank you.'*

Aldrin: *'I look forward to that very much, Sir.'*

The two astronauts now went back to their work on the moon. Armstrong's next task was to evaluate lighting and visibility.

Armstrong: *'I noticed several times in going from the sunlight into the shadow that, just as I go in, there's an additional reflection off the LM that, along with the reflection off my face into the visor, makes visibility very poor just at the transition, sunlight into the shadow. Then it takes a short while for my eyes to adapt to the lighting conditions.'*

Armstrong was now preparing to collect the main bulk of lunar samples with the aid of a long-handled scoop.

PAO: Neil's been on the surface an hour now. One hour and a half expended on the PLSS's now. Heart rates on both crew men have been averaging between ninety and a hundred. Flight surgeon reports they're right on the predicted number of thermal units expended in energy of work and he thinks they're in great shape.

While Armstrong was filling a bulk sample bag with rocks and debris, Aldrin moved out of sight to take some photographs around the lunar module.

Armstrong: '*Bulk sample is sealed* [inside one of the sample cases].' He then went over to the LM and joined Aldrin for an inspection of the craft.

Armstrong: '*I don't note any abnormalities in the LM. The pads seem to be in good shape. The primary and secondary struts* [of the landing gear] *are in good shape. Antennas are all OK. There's no evidence of problem underneath the LM due to engine fault or drainage of any kind.*'

Aldrin: '*It's very surprising, the surprising lack of penetration of all four of the foot pads. . . . On descent, both of us remarked that we could see a very large amount of very fine dust particles moving about. It was reported beforehand that we would probably see gassing from the surface after engine shutdown but, as I recall, I was unable to find that.*'

PAO: An hour and a half of lunar surface time for Neil Armstrong. They've been in the portable life support systems for two hours now.

Aldrin: '*Houston, the passive seismometer has been deployed manually.*'

The purpose of the seismometer was to detect and measure moonquakes. Inside the seismometer was a suspended weight designed to remain immobile as the rest of the equipment moved in response to tremors on the lunar surface. The relative motion was converted into an electrical impulse to give a reading of amount and frequency of movement. This electrical impulse was then to be transmitted directly to tracking stations on earth. Soon after it was deployed there was some inconclusive discussion as to whether it had detected the impact of the Soviet Luna 15 craft as it crashed into the Sea of Crises. Also mounted on the seismometer was a dust detector which was to measure the build-up of dust particles on the surface of the experiment. Power to operate the seismometer was drawn from two solar panels which converted sunlight into electrical power. Also left on the moon was the laser ranging retro-reflector experiment designed to reflect laser beams from earth and so help scientists to make extremely accurate measurements of the distance of the moon. One of the benefits of this, according to some suggestions, will be the possibility of detecting comparatively sudden changes in the moon's position as it wobbles in its orbit – a phenomenon which may have some connexion with terrestrial earthquakes. Armstrong and Aldrin had some difficulty in finding level ground on which to set up both these experimental packages.

MCC: *'We've been looking at your consumables, and you're in good shape. Subject to your concurrence, we'd like to extend the duration of the EVA fifteen minutes from nominal. Your current elapse time is two* [hours] *plus twelve* [minutes].'

Armstrong: *'OK, that sounds fine.'*

Aldrin was now collecting a core sample by hammering a tube into the surface.

Aldrin: *'I hope you're watching how hard I have to hit this into the ground to the tune of about five inches, Houston. It almost looks wet.'*

Armstrong and Aldrin were running short of time. At this stage they were supposed to take samples from specific points around the lunar module, recording carefully the exact location. In the event, Aldrin had only had time to take a few more core samples and pack up the solar wind experiment, while Armstrong picked up rocks from scattered sites as fast as he was able.

MCC: *'Buzz, this is Houston. You have approximately three minutes until you must commence EVA termination activities, over. Neil, this is Houston. Anything else you can throw into the box would be acceptable.'*

PAO: They've been on their life support systems two hours and twenty-five minutes.

Aldrin climbed the ladder back up to the hatch of the LM and crawled inside. Armstrong remained a while on the surface as they hoisted up the sample containers, the solar wind experiment and the cameras with the rope and pulley. At one point there was some doubt as to whether they had left a camera back behind.

MCC: *'Neil, this is Houston. Did you get the Hasselblad* [camera] *magazine?'*

133

Armstrong: *'Yes, I did and we've got about, I'd say twenty pounds of carefully selected, if not documented, samples.'*

Aldrin once again helped Armstrong through the hatch and into the lunar module's cabin. The first thing they then did was to transfer the umbilicals from their PLSS backpacks to the lunar module's air supply. The bulky PLSS packs and other equipment were thrown out of the hatch to offset the weight of the moon samples. This operation completed, the hatch was closed, the cabin repressurized, and the astronauts took off their helmets and gloves. Thus ended man's first brief wanderings on the surface of another world. Armstrong had spent two hours and thirty-seven minutes on the lunar surface. For Aldrin the time was some twenty minutes less. Between them they had completed almost all the planned tasks and had shown that man can live and work on the moon without any great difficulty. Dr Charles Berry, the astronauts' doctor, reported that heart rates during the moon walk ranged from a low of ninety for both crewmen to a high of about 125 for Buzz Aldrin at two periods, and a high of 160 for Neil Armstrong at three periods. The top reading came during the time he was transferring the rock boxes into the lunar module. Dr Berry said that he was working very hard at that point. Dr Berry also reported that from his readings he could not see any increase in the amount of radiation the crew were experiencing as a result of the moon walk. The success of this part of the mission was to give a great boost to the planners of future Apollo missions.

GET 114:30 (7.02 a.m. BST, 21 July) Armstrong and Aldrin were now eating a meal before starting an eight-hour rest period. During the meal the two astronauts made a further report on the area around their landing site. Armstrong said he had collected a total of fifty pounds of soil and rock samples from several areas around the lunar module. Most of them had been scooped off the surface. In some places the composition of the lunar soil was consistent down to a depth of several inches, but in others he had found hard rock beneath the soil. Aldrin confirmed that while the descent rocket engine's blast had disturbed the surface immediately beneath it, it had not made a deep crater. He also reported that the core sampler went to a depth of about three inches and then no further, indicating a solid rock bed at this depth in the area he was sampling. He repeated his impression that the compacted material in the tube looked almost moist. The report over, the astronauts attempted to get some sleep in their tiny cramped moon base cabin. Outside, the shadows shortened as the sun rose higher over the Sea of Tranquillity.

GET 121:40 (4.12 p.m. BST, 21 July) MCC: *'Tranquillity base, how's the resting standing up there?'* Aldrin: *'Neil has rigged himself a very good hammock. . . . He's been lying on the engine cover and I curled up on the floor.'* Neither of them had slept very well and now they were faced with the critical task of flying the top half of the lunar module to meet Collins in moon orbit. Completing a brief breakfast, they began a long series of checks on the spacecraft's systems, put on their helmets and gloves, and once more put themselves in the hands of pings.

GET 124:05 (6.37 p.m. BST) MCC: *'You're cleared for take-off.'* Armstrong: *'Roger, understand, we're number one on the runway.'*

PAO: This is Apollo control. We have confirmation on the ground that the ascent propulsion system propellant tanks have been pressurized.

GET 124:09 (6.41 p.m. BST) Aldrin: *'Ascent feeds are open, shut-offs are closed.'*

GET 124:11 (6.43 p.m. BST) MCC: *'Tranquillity base, less than ten minutes here. Everything looks good.'*

GET 124:18 (6.50 p.m. BST)

PAO: Eagle, you're looking good to us. We'll continue to monitor now at three minutes, twelve seconds away from ignition as the crew of Eagle goes through their pre-launch checklist. Guidance reports both navigation systems [pings and aggs] are looking good.

MCC: *'Coming up on two minutes. . . . Mark, T minus two minutes.'* Aldrin: *'Roger, guidance steering in the aggs.'* Using the lower descent stage as a launching platform, the ascent stage was poised for flight.

1. *Left:* A few seconds
after lift-off at Cape
Kennedy, the 363-foot
Saturn V on its way to
the moon.
2. *Below:* All that remains
nine days later: the
Apollo 11 command
module in the Pacific.

3. *Left:* On the launch pad at Cape Kennedy. Sitting on the mobile launcher above the flame trench, the Apollo Saturn is readied for a moon shot.

4. *Right:* Apollo 12 astronaut Richard Gordon holding a Hasselblad camera inside the command module.

5. *Below right:* A fish-eye view of Apollo 12 commander Charles Conrad and Alan Bean inside their lunar module.

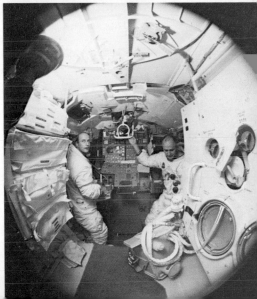

6. *Left:* Apollo 11 astronaut Edwin Aldrin climbs down to Tranquillity Base.
7. *Below:* Aldrin stands near the solar wind sample collector.

8. Apollo 12 astronaut Alan Bean prepares to deploy the lunar experiments. In front of his left foot is the miniature nuclear power pack.

9. Bean standing behind the lunar magnetometer. The lunar module can be seen in the background.

10. *Above left:* A thin slice of Apollo 11 moon rock viewed by polarized light reveals crystalline structures not unlike rocks of earth origin.
11. *Left:* A moon rock from a few inches taken by Apollo 12 astronaut Bean.
12. *Above right:* Michael Collins' view of Eagle as it was prepared for the descent to Tranquillity Base.
13. *Right:* A hastily snatched photograph of the damaged Apollo 13 service module taken a few moments after it was discarded prior to re-entry.

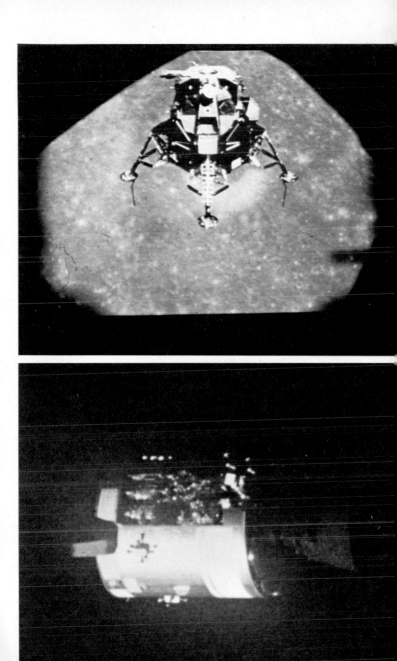

14 and 15. Artists' impressions of a space station and shuttle craft of the future.

ignition plus 16 seconds
height 650 feet,
vertical speed 50 miles an hour

ignition plus 14 seconds
height 500 feet,
vertical speed 48 miles an hour

ignition plus 10 seconds
height 240 feet,
vertical speed 34 miles an hour

ignition of ascent stage engine

LM descent stage
left at Tranquillity base

moon surface

GET 124:21 (6.53 p.m. BST, 21 July), Lift-off from Tranquillity base

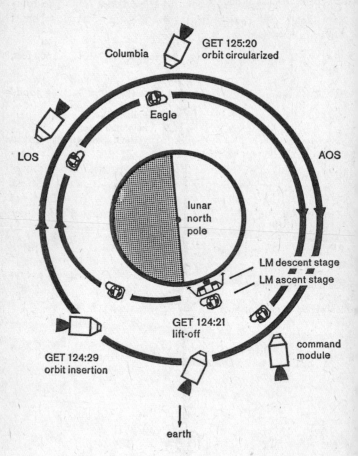

Lift-off and into lunar orbit from Tranquillity base

GET 124:21 (6.53 p.m. BST) Aldrin: *'Nine . . . eight . . . seven . . . six . . . five . . . first stage engine on ascent . . . proceed . . . beautiful . . . twenty-six . . . thirty-six feet per second up . . . little pitch over, very smooth, very quiet ride.'* The four-foot-high, 3,500-pound thrust ascent engine began its short seven-minute life.

PAO: 1,000 feet high, 80 feet per second vertical rise . . . 2,600 feet, altitude . . .

MCC: *'Eagle, Houston. One minute and you're looking good.'*

PAO: 130 feet per second vertical rise rate.

Aldrin: *'A little bit of slow wallowing back and forth. Not very much* [attitude control] *thruster activity.'* MCC: *'Roger, mighty fine.'* Aldrin: *150* [feet per second] *up . . . 9,000* [feet, altitude] *. . . aggs agrees* [with pings] *within a foot per second.'* MCC: *'Eagle, Houston, you're looking good at two. Pings, aggs and misfin* [MSFN ranging signal] *all agree.'* Aldrin: *'170* [feet per second] *up, beautiful, 14,000* [feet, altitude] *. . . within a foot per second again: aggs and pings . . . 1,500* [feet per second, horizontal velocity], *180* [feet per second, vertical velocity].' MCC: *'Eagle, Houston, you're go at three minutes. Everything's looking good.'* Armstrong: *'Roger.'*
Aldrin: *'Going right down US one.'*

PAO: Height now approaching 32,000 feet.

MCC: *'Eagle, Houston. Four minutes. You're going right down the track, everything's great.'*

PAO: Horizontal velocity approaching 2,500 feet per second.

Aldrin: *'Got Sabine to our right now.'* Sabine is a large crater on the western margin of the Sea of Tranquillity.

PAO: Some 120 miles to go to [orbital] insertion.

Aldrin: *'There's Ritter . . . that's impressive looking, isn't it.'* Ritter is a crater which touches Sabine to the north-east. MCC: *'Eagle, Houston. You're still looking mighty fine.'*

PAO: One minute to go in the burn [until orbit insertion]. 4,482 feet per second horizontal velocity.

Aldrin: '*Stand by for engine cut-off . . . shutdown . . .*' MCC: '*Eagle, roger, we copy. It's great, go.*' Aldrin: '*OK, Houston, we show 47.3 by 9.5* [the parameters of their orbit in nautical miles, given by pings]. *Aggs says 9.5* [by] *46.6* [nautical miles].' MCC: '*Eagle, Houston, copy.*' Aldrin: '*Eagle is back in orbit, having left Tranquillity base, and leaving behind a replica from our Apollo Eleven patch* [showing an eagle] *with an olive branch.*' MCC: '*Roger, we copy. The whole world is proud of you.*'

As silence fell once more on the Sea of Tranquillity, Armstrong and Aldrin were busy preparing for the next stage, a series of manoeuvres to get the ascent stage of the lunar module into the same orbit as Collins in the command module. They were in an orbit fractionally higher than intended but this was not to prove a problem. As they swung over the Ocean of Storms along the lunar equator, they prepared to fire the thrusters of the attitude control system to circularize their orbit. This would be done behind the moon when they had reached the highest point in their orbit.

GET 125:08 Loss of signal from Columbia. Two minutes later, LOS from Eagle.

GET 125:20 (7.52 p.m. BST) Eagle was now some fifty-five miles above the far-side lunar surface while Columbia was about another fifteen miles up. Eagle would have begun to drop back down towards the low point of about eleven miles in its egg-shaped orbit, but now Armstrong fired the thrusters mounted on the outside of the ascent stage of the lunar module for a duration of forty-five seconds to change its orbit to around fifty-five by fifty-five miles, just fifteen miles below Columbia's. Much of this and the following manoeuvres was in the hands of the pings computer as Armstrong fed in the appropriate programmes.

GET 125:54 (8.26 p.m. BST)

PAO: This is Apollo control. We are less than a minute from acquisition of the spacecraft Columbia coming around to the near side of the moon on its twenty-sixth revolution. We are some three minutes away from Eagle's appearance on the lunar front side. . . . We have AOS of the spacecraft. . . . Range between Columbia and

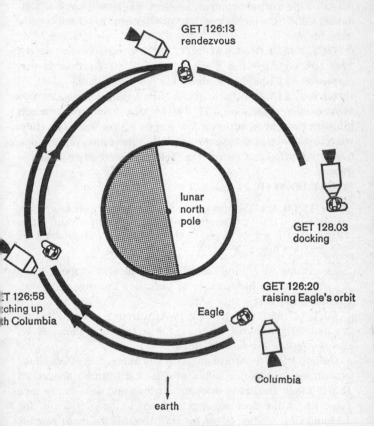

Rendezvous and docking

Eagle now showing 67·5 nautical miles [77·6 statute miles]. Range rate [closing rate], 121 feet per second.

GET 126:20 (8.52 p.m. BST) Swinging high above the landing site, Armstrong punched a computer key to fire a 1·9 second burst of the attitude control thrusters. Eagle, still some fifteen miles below Columbia's orbit, now began to climb gently towards it.

GET 126:58 (9.30 p.m. BST) A 22·4 second burn of the thrusters adjusted the flight path of Eagle so that it was approaching Columbia at about 131 feet per second.

GET 127:13 (9.45 p.m. BST) Now began a series of two mid-course corrections and five braking manoeuvres which brought the LM to within a few feet of the command module. Once again it was behind the moon that these final phases took place. At Mission Control the flight team waited to hear the results.

GET 127:51 (10.23 p.m. BST)

PAO: This is Apollo control. We are less than a minute away from acquisition of the spacecraft Columbia. Hopefully flying within a few feet of it will be Eagle. Docking should take place about ten minutes from now.

Everything was going according to plan, and as they swung around from the hidden side of the moon the two craft were preparing to dock.

GET 127:55 (10.27 p.m. BST) Armstrong: *'OK, Mike, I'll try and get positioned here. . . . I'm not doing a thing, Mike. I'm just letting her hold in attitude.'*

Collins: *'OK.'* Armstrong: *'We're all yours, Columbia.'* Armstrong and Collins could now see each other's spacecraft through their respective docking windows and were lining each other up on the docking targets affixed to the outsides of the LM and CM. Collins edged his craft towards the lunar module inch by inch. As he eased the docking probe on the nose end of the command module into the connecting ring of the lunar module, the docking latches automatically snapped shut. Collins then flipped a switch on his control panel to withdraw the probe and it was at this point that 'all hell broke loose', as Collins later remarked. The two craft began to gyrate, an event

142

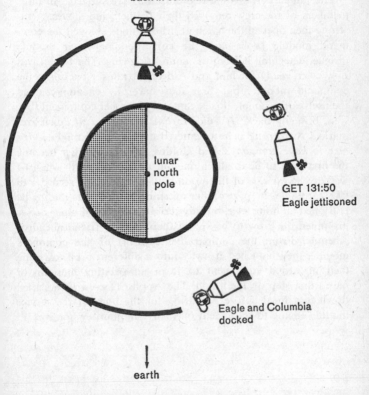

Armstrong and Aldrin
back in command module

lunar
north
pole

GET 131:50
Eagle jettisoned

Eagle and Columbia
docked

earth

Crew transfer and jettison of Eagle

reminiscent of the trouble Armstrong had with the first ever space docking on the Gemini 8 mission. Using the thrusters of the attitude control systems on both craft, Armstrong and Collins quickly brought the situation under control. Apparently, when the two craft docked, their aim was slightly inaccurate and the impact set them spinning: a dangerous moment, since there was no friction to slow them.

The tunnel between the two craft was pressurized. All three members of the crew removed their helmets and gloves. Armstrong then opened the hatch at his end and removed the command module probe and the corresponding lunar module drogue, detaching it from the connecting ring. The tunnel was now clear, ready for him and Aldrin to transfer back into the command module, but first they carefully vacuum-cleaned themselves, the sample boxes, cameras and other equipment they were bringing back. '*It's nice to find a place to sit down*,' remarked Armstrong as he re-entered the command module. '*It's nice to have company*,' said Collins, who had earlier become the first man to be totally alone out of sight of earth when he was on the far side of the moon with only his tape recorder to talk to. Eagle, its part of the mission fulfilled, was due to be jettisoned an hour later, but mysterious creaks and other noises prompted the crew to be rid of their faithful friend ahead of schedule. From the comparative comfort of the command module, they launched it away into a different orbit, carrying their discarded equipment to form an orbiting museum of man's first step on the moon. The Apollo 11 crew then settled down to a meal before preparing for the burn of the service module engine to start them on their long journey home.

6

Back to Earth
22 - 24 July 1969

As the command module began its thirty-first orbit of the moon, the Apollo 11 crew completed the final checks before the service module main engine was started up to push the craft out of moon orbit on a course for earth.

GET 135:25 (5.57 a.m. BST, 22 July) During the final orbit, some seventy miles above the far-side lunar craters, two thruster jets of the service module's attitude control system were fired for sixteen seconds against the direction of movement. This 'ullage' manoeuvre forced the fuel in the tanks towards the outlets to the engine. Then the main engine was opened up for a burn lasting two minutes twenty-nine seconds, which lifted the Apollo 11 craft high above the moon and swung it in a wide arc around to the near side and out into the spatial chasm between the moon and earth. Their speed had now built up to above the escape velocity of 5,300 miles an hour necessary to release the craft from the influence of lunar gravity. Armstrong, Aldrin and Collins were on their way home. Now they could relax, eat a meal and get ten hours' sleep, leaving themselves in the safe hands of Mission Control.

GET 142:04 (12.36 p.m. BST)

PAO: This is Apollo control. Apollo Eleven is 18,243 nautical miles [20,979 statute miles] from the moon. Velocity is 4,426 feet per second [3,010 miles an hour]. Apollo Eleven is in the passive thermal control mode. The performance of all systems is nominal [according to plan] and the crew is asleep.

Apollo 11 was rolling quietly as it carried its sleeping occupants across the gulf of space.

GET 145:00 (3.32 p.m. BST)

PAO: Apollo Eleven is 25,857 nautical miles [29,736 statute miles]

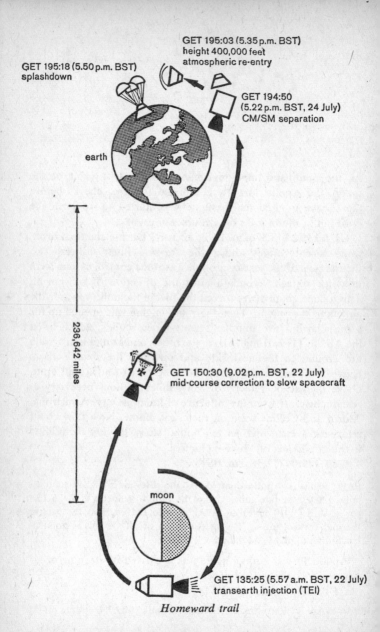

GET 195:03 (5.35 p.m. BST)
height 400,000 feet
atmospheric re-entry

GET 195:18 (5.50 p.m. BST)
splashdown

GET 194:50
(5.22 p.m. BST, 24 July)
CM/SM separation

earth

236,642 miles

GET 150:30 (9.02 p.m. BST, 22 July)
mid-course correction to slow spacecraft

moon

GET 135:25 (5.57 a.m. BST, 22 July)
transearth injection (TEI)

Homeward trail

from the moon heading towards home at 4,338 feet per second [2,950 miles an hour].

GET 146:00 (4.32 p.m. BST)

PAO: We have heard nothing from the crew. We assume they are still asleep. At this time Apollo Eleven is 28,421 nautical miles [32,684 statute miles] from the moon and travelling at a speed of 4,322 feet per second [2,939 miles an hour]. All spacecraft systems continue to look very good to us here on the ground.

GET 147:30 (6.02 p.m. BST)

PAO: The flight surgeon reported a short while ago that the lunar module pilot, Buzz Aldrin, appears to be up at this time. Both Armstrong and Collins appear still to be sleeping. At the present time Apollo Eleven is 32,253 nautical miles [37,091 statute miles] from the moon and travelling at a speed of 4,303 feet per second [2,926 miles an hour].

GET 147:39 (6.11 p.m. BST)

PAO: The flight surgeon reports all three crewmen are now awake and we expect to be hearing a good morning from the spacecraft before long.

Armstrong: *'Good afternoon, Houston, Apollo Eleven up.'*
MCC: *'Good morning, Eleven, this is Houston.'*

As they ate their breakfast, the Apollo 11 crew coasted over the point in space where the influence of earth's gravity takes over from the moon's. Their homeward speed began to build up. They were now 38,870 miles out with 200,100 miles to go and travelling at 2,716 miles an hour. Mission Control sorted out a problem with one of Aldrin's biomedical sensors and read over the day's news.

GET 150:04 (8.36 p.m. BST) The spacecraft was now lined up prior to a mid-course correction that would slow it slightly in order to bring it over the Pacific Ocean and the recovery ship *Hornet* at precisely the right moment on 24 July.

PAO: One minute now until mid-course correction number five, giving a change of velocity retrograde of 4·8 feet per second [3·3 miles an hour]. The primary purpose of this manoeuvre will be to control the spacecraft flight path angle at [atmospheric] entry interface.

The service module attitude thrusters were fired for ten and a half seconds to tweak the craft into the correct flight path for splashdown. No further corrections were needed after this one. Mission Control then discussed with Armstrong and Aldrin the problem of pin-pointing the lunar landing site, the exact position of which remained uncertain.

GET 160:00 (6.32 a.m. BST, 23 July) In the spacecraft the Apollo 11 crew ate the last meal of their 'day', a simple menu compared with the feasts now being planned by mayors of many cities back in America for the days following their release from quarantine.

GET 173:00 (7.32 p.m. BST) After another late lie-in to catch up on the two busy days around the moon, the crew were quietly eating their breakfast.

PAO: This is Apollo control at 173 hours, 18 minutes. There are virtually no flight plan activities scheduled at this time. The spacecraft systems all continue to function normally and at the present time Apollo Eleven is 106,482 nautical miles [122,454 statute miles] from the earth. The velocity is 5,607 feet per second [3,813 miles an hour].

GET 174:24 (8.56 p.m. BST) Apollo 11 was now halfway home with 118,321 miles to go. The pull of earth's gravity was now increasing their speed. Back on earth scientists were examining the first results obtained by the experimental equipment left on the moon. The seismometer, in spite of overheating, was beginning to record tremors, but the question of whether these were caused by the impact of meteors or by moonquakes was already causing a split in the scientific community. The cynical prediction that preliminary exploration of the moon would cause more controversy than it would answer questions seemed to be coming true.

GET 177:31 (12.03 a.m. BST, 24 July) Switching on their colour television camera, the crew sent back a last transmission and expressed some of their thoughts on this historic mission. Armstrong: *'A hundred years ago, Jules Verne wrote a book about a voyage to the moon. His spaceship, Columbia, took off from Florida and landed in the Pacific Ocean after completing a trip to the moon. It seems appropriate to us to* share with you some of the reflections of the crew as the modern day Columbia completes its rendezvous with the planet earth in the same Pacific Ocean tomorrow. The responsibility for this flight lies first with history and with the giants of science who preceded this effort. Next with the American people who have, through their will, indicated their desire. Next, four administrations and their congresses for implementing that will, and then to the [space] agency and industry teams that built our spacecraft: the Saturn, the Columbia, the Eagle and the little EMU [extravehicular mobility unit], the spacesuit and backpack that was our small spacecraft out on the lunar surface. We'd like to give a special thanks to all those Americans who built those spacecraft, who did the construction, designed the tests and put their hearts and all their abilities into those craft. To those people, tonight we give a special thank you, and to all the other people that are listening and watching tonight, God bless you.'

Aldrin: 'As we've been discussing the events that have taken place, in the past two or three days here on board our spacecraft, we've come to the conclusion that this has been far more than three men on a voyage to the moon. More still than the efforts of one nation. We feel that this stands as a symbol of the insatiable curiosity of all mankind to explore the unknown. Neil's statement the other day upon first setting foot on the surface of the moon: "This is a small step for a man, but a great leap for mankind," I believe sums up these feelings very nicely. We accepted the challenge of going to the moon. The acceptance of this challenge was inevitable. The relative ease with which we carried out our mission I believe is a tribute to the timeliness of that acceptance. Today I feel we are fully capable of accepting expanded roles in the exploration of space. In retrospect, we've all been particularly pleased with the call-signs we very laboriously chose for our spacecraft: Columbia and Eagle. We've been particularly pleased with the emblem of our flight, depicting the US eagle bringing the universal symbol of peace from the planet earth to the moon, that symbol being the olive branch. It was our overall crew choice to deposit a replica of this symbol on the moon. Personally, on reflecting on the events of the past several days,

149

a verse from the psalms comes to mind to me: "When I con-
sider the heavens, the work of thy fingers, the moon and the
stars which thou hast ordained; what is man that thou art
mindful of him." '

Collins: *'This trip of ours to the moon may have looked to*
you simple or easy. I'd like to assure you that that has not been
the case. The Saturn V rocket which put us into orbit is an
incredibly complicated piece of machinery, every piece of
which worked perfectly. This computer up above my head
has a thirty-eight-thousand-word vocabulary, each word of
which has been very carefully chosen to be the utmost value
to us, the crew. This switch which I have in my hand now has
over three hundred counterparts in the command module
alone. In addition to that, there are myriads of circuit breakers,
levers, rods and other associated controls. The SPS engine,
our large rocket engine on the back end of our service module,
performed flawlessly or we would have been stranded in lunar
orbit. The parachutes up above my head must work perfectly
tomorrow or we will plummet into the ocean. We have always
had confidence that all this equipment will work and work
properly, and we continue to have confidence that it will do
so for the remainder of this flight. All this is possible only
through the blood, sweat and tears of a number of people.
First, the American workmen who put these pieces of mach-
inery together at the factory. Second, the painstaking work
done by the various test teams during the assembly and the
re-test after assembly. Finally, the people at the Manned
Spacecraft Centre both in management, in mission planning
and flight control and, last but not least, in crew training.
This operation is somewhat like the periscope of a submarine,
all you see is the three of us, but beneath the surface are
thousands of others, and to all those I would like to say
thank you very much.'

GET 189:00 (12.32 a.m. BST, July 24) In the Pacific night
the final day of the Apollo 11 mission began, as a force of
9,000 men in nine ships and fifty-four aircraft took up their
stations. Around the world, nineteen tracking stations stood
by. In the recovery area some 950 miles south-west of Hono-
lulu, the aircraft carrier USS *Hornet* steamed slowly in circles

150

around the point where Armstrong, Aldrin and Collins would make contact again with their home planet.

GET 194:50 (5.22 p.m. BST) Its purpose now fulfilled, the service module was jettisoned. The command module was all that was left of Apollo 11, its weight, 12,250 pounds, less than one-fifth of a per cent of the giant tower of rocketry and electronics which had left Cape Kennedy just eight days before. Now falling towards the earth at over 24,000 miles an hour, the command module was positioned base forward so that its two-inch-thick heat shield would meet the atmosphere and the searing inferno of re-entry friction. The crew were hurtling down a narrow re-entry corridor. Aimed too high, they would have missed the atmosphere and spun off to a lonely death far out in space; too low, and the heat shield of the craft would have soon burnt through leaving the crew to their fiery end.

MCC: *'You're dead on the track, you're cleared for landing.'*

Armstrong: *'We sure appreciate that.'*

GET 195:03:27 (hours : minutes : seconds) Height 400,000 feet. A wall of hot air began to build up in front of the heat shield which, at a temperature of 5,000°F, became ionized and impenetrable to radio waves, causing total blackout of communications between spacecraft and ground for nearly four minutes.

PAO: At blackout we were showing velocity 36,237 feet per second [24,641 miles an hour]. Range to go to splash, 1,510 nautical miles [1,736 statute miles].

The spacecraft was now dropping towards the ocean at sixty miles an hour from fifty miles up, and glowing as the outer layers of its heat shield burnt away.

PAO: ARIA three reported a visual contact.

This was one of the Apollo range instrumented aircraft, operating to the west of the splashdown area. The dawn was coming up on the Pacific, the wind speed built up to eighteen knots and the sea was running a seven-foot swell. Four helicopters were busily circling around the carrier, which now had President Nixon aboard, as further reports of sightings came in. From the carrier, the fireball could be seen for a few moments through a gap in the cloud cover. Soon after the ship's radar picked up the command module as the first set of parachutes, the drogue chutes, began to brake its descent.

GET 195:13:06 The three main parachutes were now deployed at 23,000 feet.

GET 195:17:52 (5.50 p.m. BST, 24 July) Splashdown. Dragged down by the parachutes, the command module was pitching upside down in the Pacific swell, but automatically inflated floatation bags soon righted it. From one of the helicopters, frogmen jumped into the water and attached a floatation collar around the module. '*Our condition is excellent. Take your time.... It's almost as good here as in the* Hornet,' reported Collins.

The journey was over. President Kennedy's promise had been kept: '... of landing a man on the moon and returning him safely to earth.'

After securing the floatation collar around the base of Columbia, all but one of the frogmen withdrew to a dinghy tethered 100 feet upwind of the command module. The remaining swimmer, clad in a one-piece 'biological isolation garment', opened the spacecraft hatch to toss in three sets of isolation suits for the spacemen. The hatch was closed immediately. Wearing these green, light-weight, hooded suits, the crew were able to breathe in the earth's atmosphere while a biological filter prevented any moon bacteria from being exhaled. It was all part of an $8\frac{1}{2}$-million-dollar insurance policy against the possibility, however remote, of moon bugs. Recalling the decimation of many primitive or isolated earth communities upon first contact with diseases common to the 'civilized' world, NASA had been under pressure from many scientists to take steps to prevent a repetition in the space context.

The Apollo 11 crew were now ready to be received by their fellow earthmen. Crawling out of the spacecraft which had been their home for the half-million mile journey, they clambered into the navy dinghy. The hatch was immediately sealed and the exterior of the command module doused with specially prepared decontaminant chemicals. The attendant frogman, also in isolation kit, carefully scrubbed each astronaut down with the preparation, and was himself similarly treated by one

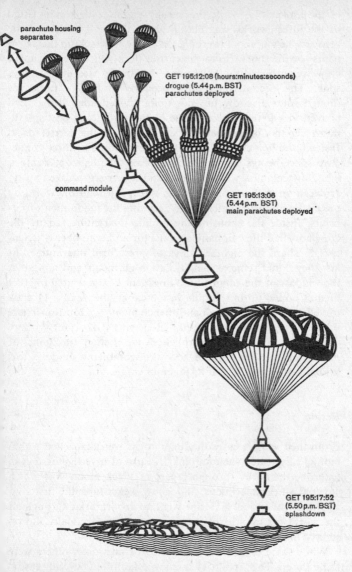

parachute housing
separates

GET 195:12:08 (hours:minutes:seconds)
drogue (5.44 p.m. BST)
parachutes deployed

command module

GET 195:13:06
(5.44 p.m. BST)
main parachutes deployed

GET 195:17:52
(5.50 p.m. BST)
splashdown

Apollo 11 splashdown, mid-Pacific, 24 July 1969

of the astronauts. While this scene, reminiscent of some ritual of the insect world, was taking place, the pilot of one of the recovery helicopters lowered a net ready to winch up the astronauts for the short trip to the carrier deck. Once the Apollo 11 crew were aboard the helicopter, the dinghy was deflated and sunk. The decontamination procedure continued, rather as though some friendly but not quite trusted alien beings had arrived in a flying saucer. None of the traditional post-splashdown ceremonial greeted the astronauts on the carrier deck. Instead, the first close view was a brief glimpse of three comic-strip space-beings stepping smartly from the helicopter into a long, low cabin. A wave, a cheer and they were whisked away; replaced on millions of television screens around the world by the men with the chemical spray, purging the moon-men's footprints. Inside the cabin of the mobile quarantine facility the crew now had the company of a doctor and a number of scientists to share the eighteen days of prescribed quarantine. At last they could remove their isolation garments and appear at the window of the cabin where President Nixon waited to greet them. Cooped up in the quarantine cabin, the Apollo 11 crew travelled by sea to Hawaii and thence by air to Houston. There they transferred to the Lunar Receiving Laboratory where quarantine conditions approached the better traditions of American comfort. There too the command module was brought, and unloaded of its precious cargo.

Results

From their sample cases the lunar rocks were unpacked within biologically sealed vacuum tanks. Teams of investigators began to probe their way through layers of black moon dust which covered all the specimens, and a few weeks later the first consignments of moon material went to laboratories throughout the world for months of intensive analysis. Gradually the results of man's first trip to the moon began to appear.

While some old questions were to be answered, others were to be raised. The Apollo 11 mission established beyond reasonable doubt that the moon is completely devoid of any life. Nor did any of the samples point to the possibility that life in any form had ever existed in the hostile lunar environment of intense radiation and extremes of temperature change between lunar day and night. Yet carbon, the elemental building block of biological molecules, was found in the lunar dust. Several explanations are offered. The carbon could have come from outside the moon, carried from the sun as a component of the solar wind of atomic particles, or it may have arrived as a constituent of carbon-rich meteorites. Alternatively it may be of 'native' origin, being the remnants of a now absent lunar atmosphere, or perhaps the result of carbon compounds outgassing from molten lunar lavas. This last suggestion is especially favoured since many of the rock samples returned by the astronauts owe their present state to having once been molten. It is now clear that the moon's history includes several phases of igneous activity resulting in giant lava flows which formed the extensive flat 'seas' of the lunar globe. Even today there are signs of such geological activity, for astronomers have noted transient glows, which may be attributed to molten rock conditions, in a crater called Aristarchus. A later Apollo mission is due to visit this site.

The age of the moon, as measured from the Apollo 11 samples, came as a surprise to those scientists who have long thought of our satellite as a younger sister of the earth. The rock samples appear to have crystallized some 3,700 million years ago while the Apollo 11 lunar dust samples are even more ancient at an age of 4,600 million years, comparable to the oldest known earth rocks. The fact that the dust is older than the rock around Tranquillity Base poses an intriguing problem for lunar geologists to ponder. If it is not crumbled local rock, what is it?

The rocks themselves appear to be made up of four common minerals: pyroxene, plagioclase, ilmenite and olivine; these same four are also common to earth rocks of igneous origin. Yet the overall chemical composition of the lunar samples is markedly different from that of terrestrial rock and adds to the evidence that the moon has long been without any atmosphere. The analyses found few of the more volatile elements normally present in solidified lavas on earth. In the lunar

155

vacuum these had boiled away when the original material was in a molten state. Any atmospheric pressure would have prevented or slowed down this process. Moon rock revealed only three new minerals previously unknown to geological science. None radically different to earthly compositions, their distinction depending on minor details. More exciting to the investigators however were the now famous 'moon marbles'; tiny glassy beads found among the samples. Whether they were formed when meteorites crashed into the moon or whether they are simply the result of lunar volcanic activity remains to be seen. The age-old controversy between those who believe the moon owes its present surface features to meteoritic impact and others who maintain that the moonscape is of volcanic origin remains as fierce as ever. Yet the Apollo missions are perhaps beginning to make some scientists wonder whether both theories may not be valid.

The origin of the moon remains just as much a mystery as before Apollo 11. Perhaps Darwin's idea that the moon was born out of the Pacific Ocean has lost its few remaining followers, but three other theories remain. The first of these suggests that the moon is but a droplet left over when a giant molten planet of the sun split up to form the earth and Mars. A second suggestion envisages the moon as a small previously independent planet of the sun that has somehow been subsequently captured by the earth's gravitational field. The third theory, currently the most fashionable, postulates that the moon is the result of the coagulation of many much smaller satellites of the earth. Originally these 'planetismals' circled the earth much as Saturn's rings do today. This proposal has many attractive qualities and would help to explain the moon's pock-marked appearance, its 'mascons' or massive concentrations of denser material below the surface of various parts of the moon, and several other features that are not so easy to explain in terms of the rival theories. The Apollo missions have become the chief talking-point whenever geologists forgather.

From a less scientific point of view the Apollo 11 mission can be said to have vindicated those who had constantly argued in favour of manned, in preference to unmanned, space exploration. Though guided by sophisticated electronics, Armstrong was able to take over the controls of Eagle when the lunar module computer became overloaded and to guide the craft away from a boulder-filled crater to the safety of Tranquillity Base. Without a man on board, the craft would have continued its perilous course to end up the same way as Luna 15.

Luna 15's demise also underlined the radically different approach in the Soviet Union to space exploration. Shrouded in secrecy and dominated by senior Academicians, it appeared in sharp contrast to the American space effort. The latter was the more risky, especially when carried out in full view of the whole world, yet produced bigger results for the same investment of money and men.

14 November 1969 saw the launch of the second Apollo moon landing mission. All three Apollo 12 astronauts were Navy men. Charles (Pete) Conrad commanded the mission, with Richard Gordon as command module pilot and Alan Bean as lunar module pilot. Preparations for the launch went as smoothly as those for Apollo 11 until two days before lift-off. A component of the fuel cell system aboard the service module showed signs of a failure and had to be replaced hurriedly with the respective part from the Apollo 13 service module, which was already at Cape Kennedy.

The launch itself was a spectacular. A rain-storm had closed in on the Cape area a few hours before lift-off. While the remaining minutes ticked by a cloud totally obscured the Saturn V from the launch control centre. Under normal circumstances this would have been enough to postpone the shot, even though it would have meant a month's delay to the mission, but on this day NASA had a special guest. For the first time a president had come to see a manned launch at the Cape, and NASA, especially sensitive to its image in the eyes of its paymaster-general, was not going to disappoint President Nixon.

Exactly on time, nine seconds before lift-off, the Saturn engines were started up. From the launch control area the cloud over the pad was seen to fill with an orange glow. As though by some trick of cinematography the shiny white Saturn appeared through the swirling vapours. Rising on five tongues of red fire it seemed to float momentarily and then disappear. The whole sequence of events took place in muted silence as observers some three and a half miles back held their collective breath. Then, as the sound of roaring engines reached the launch control area, two forks of blue-white lightning streaked down through the cloud from either side of the Saturn's path. Over the communications circuit came the voice of Conrad: *'OK, we just lost the platform, gang. I don't know what happened here. We had everything in the world drop out ... fuel cell, lights, AC ...'* Even the telemetry shut off, so for a few anxious moments Conrad's voice was the only link with the spacecraft soaring out over the Atlantic. With electric power off, the command module's automatic guidance system was no longer functional. Almost as a natural reflex Conrad switched over to reserve power from batteries. For over a minute things looked none too well but gradually the power supply was restored. The vital computers were undamaged and the second-stage engines of the Saturn lit up exactly on schedule. Only part of the navigation electronics had sustained any lasting effects but these were ironed out later as Apollo 12 began its first orbit over Africa. It was at first feared that the Saturn had been struck by lightning. Later it was concluded that with thunder about the launcher had picked up a giant charge of static electricity which discharged from 600 feet up, down the ionized exhaust plume of the first-stage engines. In any event embarrassed NASA officials explained at a press conference afterwards that there were to be no more launches through thunder clouds. With this crisis past the journey to the moon went more peacefully. With a laconic reference to all-weather flying the Navy pilots went on their way towards the lunar equator for a landing at an apt choice of sites: the Ocean of Storms.

By 19 November (BST) the Apollo 12 craft was in moon orbit. Conrad and Bean on board the lunar module Intrepid were to make the landing, Gordon was to remain in orbit aboard the command module Yankee Clipper. Following the undocking of the two craft, Gordon fired the CSM attitude control jets so as to place himself some two and a half miles from Intrepid at the moment when the LM descent engine was to be started up.

GET 109:24 Conrad and Bean were now 'go' for the landing. A computer-controlled 28.2 second burst from the LM's descent engine reduced its speed by forty-nine miles an hour. Intrepid began to fall towards the moon, flying feet first with the crew facing up and away from the moonscape.

GET 110:21 Intrepid was now 50,000 feet above the approach to the landing area. Again the computer acted to guide the craft through the last 270 nautical miles of the descent. The engine started up once more to bring the spacecraft to a position 500 feet above the landing zone. Only now did Conrad and Bean take over the flying. While Conrad handled the controls, Bean called out the instrument readings: '*190 feet ... Come on down ... 180 feet ... 9 per cent* [remaining fuel], *you're looking good. Gonna get some dust before long ... 130 feet ... 124 feet, Pete ... 120 feet, coming down in 6* [feet per second] *... 8 per cent* [fuel] *... You're looking OK ... 96 feet, coming down in 6* [feet per second] *... Slow down the descent rate ... 80 feet, coming down in 4* [feet per second] *... You're looking good ... 70 feet, looking real good ... 63 feet ... 60 feet, coming in 3* [feet per second] *... 50 feet, coming down, watch for the dust ... about 46* [feet], *low level ... 42 feet, coming down in 3* [feet per second], *... coming down in 2* [feet per second], *OK. Start the clock ... looking good, watch the dust ... 32, 31, 30 feet ... coming down in 2* [feet per second]. *Pete, you got plenty of gas, plenty of gas babe, stay in there ... 28 feet, coming down in 2* [feet per second]. *He's got it made. Come on in there ... Contact light!*'

A cloud of lunar dust had obscured the landing surface, so Conrad used the light signal generated by the contact probes as the cue to cut off the descent engine. These probes extended about five feet beyond the pads of the landing gear. At this point the craft was almost standing on the exhaust jet of the descent engine which added to the gentleness of the touch down, at GET 110:32.

Conrad: '*I think I did something I said I'd never do. I believe I shut that beauty off in the air before touchdown ... Well I was on the gauges. That's the only way I could see where I was going.*'

MCC: '*Roger Pete. The Air Force guys say that's a typical Navy landing.*'

Conrad and Bean were to spend nearly eight hours exploring the lunar surface on foot in two separate EVAs. At GET 115:23 with a high-spirited '*whoopee!*' Conrad leaped from the lunar-module ladder on to the moon for the start of the first

160

EVA. Referring to Neil Armstrong's first words on stepping on to the moon, he added: *'Man, that may have been a small one for Neil, but that's a long one for me.'* As Conrad placed both feet on the ground, Gordon was passing overhead in Yankee Clipper and spotted both the LM and the nearby Surveyor through the craft's sextant. Conrad now reported to Bean: *'I can walk pretty well, Al, but I've got to take it easy and watch what I'm doing. But you'll never believe it. Guess what I see sitting on the side of the crater: the old Surveyor.'* They had landed just north-east of the crater in which Surveyor 3 had come to rest thirty-one months before. This was the target of the Apollo 12 mission and they had landed within 600 feet of it. From inside the LM Bean was watching Conrad's progress in one-sixth gravity. *'Your boots are digging in the soil quite a bit,'* he told Conrad. *'Your left foot has a big mound ahead of it right now just pushing it along.'* The cohesive black dust was to prove something of a problem throughout the two moon walks. Both astronauts and equipment became repeatedly covered in it. It was as if they were exploring a coal tip. *'I'm getting dirty,'* reported Conrad.

Adjusted to life under moon gravity conditions Conrad set about collecting the contingency sample. For this he scooped up a bagful of soil. Next he hooked up the bag to the rope and pulley system of the lunar equipment conveyor (LEC) to pass the sample up to Bean in the LM. Unlike the business-like exchanges between the Apollo 11 moon walkers, Conrad and Bean conveyed a greater impression of excitement and enjoyment as they went about the moon. High-pitched laughs and even a jaunty *'dum, dum, dee, dum,'* came down the link to Houston.

Twenty minutes into the moon walk an unfortunate accident terminated live colour television coverage of the event. While moving the camera to a new vantage point, Bean accidentally pointed it at the sun which burnt out the sensitive retina. Having spent a few valuable moments trying to get an image back the two astronauts went on to set up an umbrella-like antenna and point it towards earth to provide improved S-band communications. Conrad then unfolded a strip of special foil to collect a sample of solar wind. An American flag was planted and the two astronauts began to unpack the experimental packages that they were to leave behind on the Ocean of Storms. Providing power for these was a miniature nuclear reactor, the radio-isotope thermoelectric generator (RTG). Conrad and Bean experienced some difficulties in getting its small plutonium fuel element out of its case but eventually all the experiments were in place some 600 feet from the LM. Three of them began sending data back to earth at once including the seismometer which recorded the astronauts' footfalls as they worked around it. Other experiments were to detect and measure charged particles and magnetic properties of the lunar environment. While deploying this equipment the two astronauts noted the presence of several four to five foot cone-like mounds in the vicinity. These and many other geological features were reported by the two men who lamented the short duration of their lunar stay. Again and again they commented on the dust problem. '*We are really getting dirty out here. There's no way to handle all this equipment with all the dust on it. Every time you move something, the dust flies and in this low gravity really takes off, goes way up and it comes in and lands on you . . .*' reported Conrad. As they were unpacking one of the lunar surface experiments, Bean tossed a piece of plastic packing-foam up into the lunar sky and marvelled as it went on climbing, under one-sixth gravity and no air resistance, to a height of 300 feet. The time had now come to collect more samples and take photographs of the landing site's many intriguing features. On the way back to the LM they began to trot, estimating that they could make strides of up to ten feet with ease. Now and again they paused to select further rock specimens and to drill a sixteen-inch core through the surface. Brushing off as much of the lunar dust as possible, they climbed back into the LM at GET 118 : 50 to complete man's second moon walk.

Down in the Houston control-room Neil Armstrong expressed surprise at the number of tasks Conrad and Bean had accomplished in the time available. During the EVA the heart rates of both astronauts, as monitored by the biomedical teams at Mission Control had been remarkably low on average and neither man had needed the full supply of water in his backpack to cool his suit during the heavier work phases. At the end of the first trip outside they still had sufficient life-support consumables for two more hours. This was encouraging news for the planners of later Apollo missions with extended lunar exploration in view.

Twelve hours later Conrad and Bean were back on the lunar surface. Severing the inert television camera from its cable, they packed it for return to earth. This unexpected extra piece of luggage meant however that thirteen pounds of rock samples collected during the first EVA had to be discarded. The astronauts then walked to the site where they had previously deployed the experiments. To test the sensitivity of the seismometer Conrad rolled a rock into the 400 foot diameter Head Crater to the south of the experimental area, while scientists at Houston watched the traces relayed from the instrument. Conrad and Bean now moved south along the western edge of this crater to the northern edge of the 100 foot diameter Bench Crater. From here they moved west to Sharp Crater, some 500 yards south-west of Intrepid. Circling the crater edge they headed east again for the Surveyor crater, collecting rock and core samples and taking photographs as they progressed across the dusty surface, pock-marked with shallow craters of every size.

Surveyor 3 stood about half-way down the eastern slope of a 600 foot diameter crater to the south-east of Intrepid. The astronauts approached the southern rim. They found the gentle slope to be firm ground rather than loose rock so the going for their first venture into a sizable lunar crater was easy. They were surprised to note that the Surveyor, mostly white when launched, was now brownish in colour. The craft was covered in fine dust. Conrad and Bean removed pieces from the Surveyor in order to show its designers how the various components had stood up to two and a half years in the lunar environment. Climbing out of the crater the astronauts went north and then west again to the landing zone. Having retrieved the solar wind experiment, they prepared to load Intrepid with the results of the mile-long moon walk. With Bean up in the LM cabin, Conrad waited below to pass up the sample containers, cameras and other equipment. '*I feel just like a guy at the shopping-centre with the groceries waiting for his wife,*' he remarked.

The trip back to earth with the precious scientific cargo was routine. Only on splashing into the Pacific Ocean on 24 November did a minor incident mar an otherwise happy ending. The command module landed in the upright position but then a wave rolled it over. Stabilization bags, inflating automatically, soon righted the craft but not before a camera fell from its bracket and struck Alan Bean above an eye. Seventy minutes later the Apollo 12 crew were aboard the carrier Hornet, and Bean was given two stitches by the quarantine-facility doctors.

From both scientific and engineering points of view the success of the Apollo 12 mission was a great encouragement to NASA. For moon geologists, the crew had carried back 108 pounds of carefully documented rock and dirt samples; over fifteen pounds of random rock specimens and core samples, as well as some fifteen pounds of Surveyor equipment. They had taken hundreds of photographs over a wide area of the Ocean of Storms. Gordon, circling above in the command module had added to the riches of new information about the moon by conducting a major photographic survey of future landing sites. Long after the Apollo 12 crew had departed the moon, the experiments left behind continued to feed back data to the scientists at Houston. Much more advanced than the rather rudimentary apparatus deposited in the Sea of Tranquillity by the Apollo 11 mission, they did much to mollify the criticism that not enough science was being carried out on Apollo missions. One of the experiments, the lunar ionosphere detector, was designed to measure the flux, density, velocity and energy of positively charged ions in the vicinity of the lunar surface. It was sensitive enough to measure the increase in such particles when the ascent stage of Intrepid was fired to begin the journey back home. The solar wind spectrometer, left to measure the strength, velocity and direction of electrons and protons coming from the sun, gave scientists an added bonus when they were able to detect the influence of the earth's magnetic field on this radiation. The moonquake recorder, or passive seismic experiment, also functioned well. As well as measuring the effects of the impact when the ascent stage of Intrepid was crashed on to the lunar surface it was soon fascinating scientists with other shock waves reverberating beneath the moon's surface. Though the results are presently puzzling many scientists, it is experimentation like this that will lead to a better understanding of the nature of the moon's interior. The lunar surface magnetometer supplemented the work being done by the solar wind experiment as well as providing details about the weak magnetic forces that influence the lunar environment. Another experiment similar to the solar wind collection device was also returned to earth. This was designed to trap particles from the cosmic radiation that pervades all of space.

While Apollo 11 had proved that man can land on the moon and return safely, Apollo 12 showed how this capability could be expanded into a major scientific and exploratory advance. Following Apollo 12 both astronauts and scientists were clamouring for NASA to plan for considerably extended periods of lunar exploration for future Apollo missions. The ease with which Conrad and Bean had frolicked around on the moon lent NASA a new confidence. Apollo 13 was to be an even bolder venture.

At thirteen minutes past the thirteenth hour (local time) on 11 April 1970, Apollo 13 left the launch pad at Cape Kennedy minus one of its original crew members. Following exposure of the three astronauts to German measles, the command module designate, Thomas Mattingly, failed to show immunity to the disease and was grounded just twenty-four hours before the lift-off. His place was taken by a member of the back-up Apollo 13 astronaut pool who had been training alongside the prime crew. John Swigert was also by good fortune a civilian engineering test-pilot whose chosen speciality was responses to malfunctions in the command and service modules. Commanding the mission was James Lovell, veteran of three space flights including Apollo 8. Fred Haise, a newcomer, acted as lunar module pilot.

The launch itself was not without hazard. High above the Cape the centre engine of the Saturn V second stage shut down prematurely. A vibration in the J2 engine resulted in reduced flow through the fuel pump. Safety switches cut out the engine. Consequently the remaining four engines of the stage were kept firing for longer than normal and it took an extra long burst from the third-stage engine to put Apollo 13 into earth parking orbit.

The first signs of serious trouble were noted by the automatic telemetry being passed down to tracking stations on earth on 13 April. The mission had been out for some forty-six hours and was well on its way to the Fra Mauro uplands near the lunar equator. The spaceship now consisted of the three modules linked together (see diagram on page 44). The Saturn third stage, though also bound for the moon, had been separated following the transposition and docking manoeuvre.

Apollo 13 was not travelling along a 'free return' trajectory at this stage. This type of trajectory, a feature of earlier Apollo moon missions, would have carried the spaceship on a loop around the moon and, making use of lunar gravity, the craft would have swung back on to a return course to earth. Thus if anything went wrong, the astronauts on earlier moon shots would have just had to sit tight. Confidence in the performance of the spacecraft following Apollo 11 had led to the abandonment of this procedure. Apollo 13 was on a course less costly in fuel but greater in risk. Without a major engine-firing they were destined for orbit around the sun. This factor was to add to the perils Apollo 13 was about to face.

GET 46:40 A gauge measuring the amount of liquid oxygen in tank No. 2, went off the top end of its scale. This indicated, in theory, that the tank contained more than 100 per cent fuel load. Something was evidently amiss with either the tank or the gauge. For the time being Houston reckoned it was a fault with the gauge. No action was taken. (The two liquid-oxygen tanks aboard a service module feed a breathable atmosphere to the crew and power the three fuel cells which generate both electrical power and drinking-water. Fluctuations in the pressure of the liquid would result in fluctuations in voltage output; such pressure changes could be caused by variations in the temperature of the liquid, so fans within the tanks are switched on at intervals throughout the mission to ensure an even temperature.)

GET 47:54 As part of the normal routine the fans were switched on to stir the oxygen.

GET 55:00 The crew began a television transmission back to earth.

GET 55:42 Pressure gauges were indicating normal conditions in both tanks: 920 psi (pounds per square inch) in No. 1, 924 psi in No. 2. These tanks, situated only a few feet behind the astronauts as they lay in the command-module couches, were built of two walls and were designed to withstand pressures of above 1,000 psi, at which point a relief valve would open. The two tanks were interconnected.

GET 55:48 The alternating current coming from bus B, a distributer bringing power from the fuel cells, registered a short sharp jump in voltage. Haise noted the nine volt spike but did not mention it at this stage.

GET 55:52:05 (hours:minutes:seconds) Pressure gauge in tank No. 2 now indicated a drop to 891 psi. Houston requested the crew to recycle the stirring fans.

GET 55:53:18 Pressure in tank No. 1 down to 887 psi.

GET 55:53:29 Fans began to stir contents of both tanks.

GET 55:53:38 The meter monitoring the voltage showed a nine volt drop.

GET 55:53:42 Pressure in tank No. 2 up to 899 psi.

GET 55:54:28 Pressure in tank No. 2 up to 975 psi. A caution and warning light should have come on in the command module. It did not.

GET 55:54:29 Quantity gauge showed only 6·6 per cent oxygen remaining in tank No. 2. This indicated that the tank was almost empty.

GET 55:54:30 Temperature gauges now showed the liquid oxygen to be heating up.

GET 55:54:31 Quantity gauge now read 74·5 per cent for tank No. 2. Mysteriously this was the normal reading. However other readings were not returning to acceptable levels.

GET 55:54:34 Temperature gauges registered a further increase.

GET 55:54:35 Tank No. 2 pressure and temperature building up. Quantity gauge still steady at 74·5 per cent.

GET 55:54:37 Pressure in tank No. 2 was 987 psi. Three seconds later it had reached 996 psi. Liquid oxygen was boiling away. Pressure and temperature showed a rapid rise. Quantity gauge began to fall. A meter recording flow between tank No. 2 and fuel cell No. 3 was now oscillating wildly.

GET 55:54:43 Pressure in tank No. 2 reached 1,004 psi. Relief valve should have opened. For nine seconds the pressure stays above 1,000 psi.

GET 55:54:52 Tank No. 2 or its associated tubing burst. The explosion was heard in the cabin. It was somewhat muted owing to the absence of air to carry the shockwave. Pressure in the tank dropped sharply. Tank No. 1 is also losing pressure.

Four seconds later the master caution light flashed in front of Haise on the command module console. A warning tone sounded in all three crew members' head sets. Twenty-four seconds later came the first report of the crisis to Houston: '*Hey, we've got a problem here.*' The crew had just completed their television transmission and were discussing the photography of a comet. A skeleton corps of controllers were manning the Mission Control consoles. Only a handful of journalists was present in the vast Houston press area. The American television networks, judging that the public had grown bored of the routines of moon travel, had ignored the television from space. Ironically one station was showing for preference a space fiction programme at the very moment when the real counterpart met its unhappy fate.

MCC: '*This is Houston. Say again please.*'

Lovell: '*Houston, we've had a problem. We've had a main bus B interval* [referring to the power drop from the fuel cells].

MCC: '*Roger. Main B interval. OK. Standby, 13; we're looking at it.*'

Haise: '*OK. Right now, Houston, the voltage is looking good . . . We had a pretty large bang associated with the caution and warning* [light] *there. And, if I recall, main B was the one that had an amp spike on it once before* [referring to the power surge at GET 55:48].'

MCC: '*Roger, Fred.*'

In rapid succession the crew reported further on the various aspects of the trouble. Two of the three fuel cells aboard the service module were showing signs of deterioration. Without active fuel cells the command module relied solely on its own batteries which in turn were normally recharged from the fuel cells. With one fuel cell out of action the moon landing was automatically cancelled. With more than one fuel cell lost the mission was faced with disaster.

GET 56:12 Over the public affairs circuit from Houston came this terse report:

This is mission control. The rapid exchange of conversation you've heard . . . the main B bus is off the line; fuel cells one and three also off the line; fuel cell two is presently on the line. We now show 13 at an altitude of 178,643 nautical miles.

The crew now reported that they could see gases venting from

169

the service module through the left-hand side window of the command module. The explosion had blown an entire thirteen-foot-long panel out of the side of the service module and debris was also beginning to float past the command module windows. The whole spacecraft began to tumble and roll. Swigert: *'Something is giving us reach, both in pitch and roll ... I had to use direct* [manual over-ride control] *in order to stabilize it.'*

MCC: *'Standby. What direction are your rates* [of pitch and roll] *in, Jack?'* Swigert: *'It's negative pitch and negative roll.'* Later the pitch and roll became positive. Swigert struggled to stabilize the craft: *'I've got a positive pitch and I can't stop it.'* MCC: *'OK, OK, Jack. Bring A-4* [attitude control jets] *on, stop the pitch rate.'* Swigert: *'OK. That got it.'*

By now a tension was discernable in the voices of the otherwise steely-nerved astronauts. Each time Houston broke off the exchange there was an anxious call from the spacecraft. They clearly found comfort in the level tones of Mission Control: *'OK, 13. We've got lots and lots of people working on this ...'* Telephones were ringing all over Houston and men were rushing into the Manned Spacecraft Centre from all over the United States.

The time had now come for a series of grave and irreversible decisions. With two of the three fuel cells no longer providing electricity, the command module would soon be short of power. The loss of oxygen was also of prime concern. Efforts were now made to conserve the oxygen remaining in tank No. 1. Attempts were made to isolate the leak using the various valves in the complex of tubing that linked the tank to the fuel cells and command module cabin. This measure failed leaving one option open. One by one the vital fuel cells were closed down in a last desperate move to prevent the oxygen loss. Once shut down they could not be reactivated. In a matter of minutes all three cells were dead. Life in the command module now depended on limited supplies of oxygen, drinking-water and battery power. There were sufficient consumables for only a few hours, not enough to keep the three astronauts alive for the three day return-journey ahead of them. In any case the oxygen and battery power were vital to the last phases of the trip when the command module was to be separated from the other modules prior to splashdown. For once luck was on the side of the Apollo 13 mission. The lunar module was still attached and provisioned with its full load of fuel, oxygen, water and power. With the service module now little more than a dead hulk, the lunar module took on the combined role of the command and service modules. Safely within the lunar module cabin, the crew powered down the command module.

For the next eighty-four hours the thoroughness of the Apollo philosophy of flexibility was to face its keenest test. Not specifically designed for its new function, the lunar module was nevertheless a spaceship and as such could be called on to protect and harbour a crew facing the loss of their mother ship. Furthermore it was equipped with two large engines in its descent and ascent stages that could now be used to make the necessary course changes to bring the astronauts safely home. The lunar module's onboard navigation and computer equipment could also be used to conserve the command module's dwindling power supplies. However imminent the Apollo 13 crew's demise may have appeared at the time, NASA was from now on fully in control of the situation. The years of preparation for Apollo were now paying off. NASA now turned to its miles upon miles of computer memory. At the Massachusetts Institute of Technology (MIT) scientists looked out their 'what would happen if' tapes, and at both Houston and Cape Kennedy experienced astronauts began 'flying' the training simulators in the new configuration. One by one the answers emerged. For a start it was decided not to jettison the defunct service module. Although this would have reduced the spacecraft combination weight and saved rocket propellant it would also have meant exposing the command module heat shield to the unknown rigours of three days in space. Turning off the command module's power, and thus its guidance and navigation system, seemed at first to mean running the risk that it might deteriorate under low temperature conditions before it could be switched on again just prior to re-entry. MIT computers indicated that this risk would be slight.

GET 61:29 The first of four engine burns was now made with the lunar module's descent rocket motor. This placed the crippled ship back into a free return trajectory. Rounding the

171

moon at an altitude of 136 nautical miles, the crew began to feel the friendly pull of earth's gravity.

GET 79:28 Apollo 13 was more than 5,000 nautical miles along its 250,000 miles journey home when a second engine burn was made to shorten the journey by ten hours and bring it to a conclusion in the main recovery area in the Pacific. Had this burn not been made the craft would have been targeted for the Indian Ocean where the US Navy's presence was sparse, though offers of assistance had by now been received by navies of other nations including the Soviet Union.

For the long trek homewards the crew lived on restricted water, oxygen and power. There was plenty of food but rest periods presented a problem. At first one or two astronauts used the unheated command module as a dormitory but it had grown too cold and sleep was impossible. Later they made do with the tunnel linking the two craft but only one astronaut could sleep there at the same time. As the atmosphere became stale the crew managed to rig a purifying device making use of components of the command module's environment control system connected together with tubing from the moon-walking backpacks.

GET 105:18 (15 April) A third main engine burn was now made to steepen their angle towards the earth. For this manoeuvre the crew had to align the craft by sighting the sun and the earth's horizon. The automatic equipment had been switched off to conserve power. This burn was only partly successful.

GET 137:39 (17 April) A fourth and final course correction was still necessary to aim the stricken craft down the critical re-entry corridor. A twenty-three-second burst from the attitude control jets of the lunar module ended any anxiety about their course.

GET 138:02 Cameras at the ready, the crew jettisoned the damaged service module. Lovell reported: '*There is one whole side of the thing missing ... Looks like a lot of debris hanging out the side near the S-band antenna.*' The missing thirteen-by-five-foot-six-inch panel revealed the full extent of the havoc caused by the burst oxygen tank: it was perhaps as well for the morale of the crew that they had not seen this before. The three tired astronauts then scrambled back into the command module and one by one the batteries were turned on. All was well. Odyssey had survived the three days of enforced retirement.

GET 141:30 *'Farewell, Aquarius, we thank you.'* Lovell's last tribute came as the lunar module was separated. Odyssey was on her own and the perilous journey was almost over. The astronauts' spirits rose as the surface of the earth began to fill their porthole windows. An hour and twenty-four minutes later the three giant parachutes lowered the singed command module in the warm waters of the Pacific, four miles from the recovery ship USS Iwo Jima. The mission of Apollo 13 was mercifully over.

In spite of the fact that the damaged Apollo 13 service module had been burned up in the earth's atmosphere on re-entry, a full-scale inquiry mounted by NASA was able to pin-point the cause of the rupture of the oxygen system of the fuel cells which nearly terminated the mission in disaster. The fault lay in the use of inadequate thermal switches. These had been used during pre-launch tests to turn on the tank heaters in order to empty these containers of the oxygen. The voltage used by the ground crew had damaged them and when they were activated subsequently during the flight, a short circuit occurred which started a fire within the damaged insulation of oxygen tank No. 2 and led to its rupture. Apollo 14 was immediately delayed while modifications were made. Fireproof materials were introduced into this sector of the service module. A third oxygen tank and a 400 ampere-hour battery were installed in the SM to provide an additional source of power in case all three fuel cells should fail. Provision was also made to store an extra twenty pounds of drinking-water on the command module.

The failure of Apollo 13 enhanced the achievements of Apollos 11 and 12. But it also firmly deflated the optimistic confidence that had been growing in NASA and suddenly retarded prospects for the scientific exploration of space. It had taken three highly experienced test-pilots to get the crippled ship home: it would be a long time before NASA would risk flying 'passengers'. For the scientists training to become astronauts, Apollo 13 meant a greatly lengthened wait.

Apollos 14 to 17

As well as the setback caused by the failure of the Apollo 13 mission, NASA was also faced in 1970 with a severe clipping of its economic wings. During the year many scientists and engineers were laid off while a new set of plans was drawn up. Apollo was now to run to Apollo 17, not Apollo 20 as originally planned. Apollos 14 and 15 were to fly in 1971, and the remaining two flights would be completed in 1972. The Americans would then give the moon a rest. In late 1972 or early 1973 one of the Saturn Vs released from the Apollo programme would be used to launch Skylab, the first experiment towards the setting-up of a permanently manned earth-orbiting American space platform. But, in spite of the Apollo 13 mishap, the remaining four Apollo missions were to represent a new phase in lunar exploration. More ambitious landing sites would now be visited.

Apollo 14, due to be launched on 13 January 1971, with Alan Shepard, veteran of America's first manned launch, Stuart Roosa and Edgar Mitchell, is targeted for Apollo 13's landing site in the hilly region some fifteen miles north of the Fra Mauro crater. For the geologists, the crew hopes to bring back examples of rock material dredged up from deep inside the moon when a smaller moon or large meteorite impacted into the region some four to five billion years ago. It is hoped that, from such material, scientists will be able to unravel the early history of the moon, the earth and of our solar system, a period difficult to study in the record of rocks here on earth due to the millions of years of erosion within our corrosive atmosphere. The Apollo 14 crew is also due to conduct seismic experiments, sending vibrations through the lunar crust by detonating small explosive charges. A laser reflector is to be set up on the moon to enable observers back on earth to measure the relative wobble of the two celestial bodies which may, in the view of some optimistic experimenters, enable us to predict earthquakes much as we now forecast the weather.

Apollo 15, provisionally planned for launching on 25 July 1971, has David Scott, Neil Armstrong's companion on the hazardous Gemini 8 flight, earmarked as commander with two newcomers Alfred Worden and James Irwin as command-module pilot and lunar-module pilot respectively. This mission is targeted for a northern lunar plane cut by a large gorge that runs along the base of some of the moon's highest mountains. The site has been named Hadley-Apennine and the 600-foot-deep gorge, not unlike a dried-up river canyon, is called the Hadley Rille. It is hoped that the astronauts will be able to collect samples of rock from the base of the moon's Apennine mountains, rising nearly 8,000 feet above their landing site. As in the case of the Apollo 14 samples, it is hoped that these rocks may have been ejected from deep within the moon following the impact which created the moon's Imbrium basin. The crew will also explore some of the 60-mile-long rille. This, it is suggested, may have cut into the geological sandwich of lunar history in much the same way as the Grand Canyon has on earth. The Apollo 15 crew are to be sped on their travels by a two-man electric cart to be carried in the descent stage of the lunar module. While two of the crew are exploring the rille, the third man in lunar orbit will be conducting an extensive photographic survey. It is also planned to launch a small lunar satellite from the service module which, among other things, should provide more information about the mysterious lunar mascons. Another innovation on the Apollo 15 mission will be a longer stay-time on the lunar surface. Starting with this flight, a modified spacecraft will be flown allowing for a sixty-six hours' period to be spent on the moon.

Apollos 16 and 17, due to fly in 1972, are, under the new plan, to complete the Apollo saga. NASA will then turn to the main project of the 1970s, the construction of a large space station in earth orbit supplied by a re-usable shuttle rocket.

This programme, announced in late 1970, is considerably more modest than that outlined in May 1969. Then, following months of deliberation, the members of a NASA committee set up to preview the 1975–85 period had outlined in their final report the areas of manned space exploration recommended for further development. In summary they called for extensive exploration of the moon to be closely associated with semi-permanent and eventually permanent lunar colonies. Further afield the committee called for the exploration of Mars, one of our next-door neighbours in the solar system, initially with un-manned orbiting and landing craft but later with a fully fledged manned expedition. It was also proposed that large manned earth-orbiting space stations should be constructed, particularly for the purpose of studying many resources of our planet that remain as yet untapped or poorly exploited. These giant space stations would also serve as observation posts and laboratories for scientists of many disciplines and very probably the military. A USAF project to launch a manned orbiting laboratory which had already cost the American taxpayers $1,600 million was dropped, in exchange, it must be assumed, for a military presence aboard the NASA space platform.

The cost of these ambitious space ventures worked out at around ten billion dollars a year, or about half a per cent of America's annual gross national product. This was about twice NASA's largest annual budget and assumed that America's wealth would continue to grow through the seventies and eighties at the present rate. These estimates were presented in mid 1969 when the euphoria of Apollo 11 was at its height and when even Vice-President Agnew was talking of an American on Mars by 1980. Few seemed to notice that this kind of money was not there nor would it be in the forseeable future. The big race was over. Other national priorities cut drastically into the space budget. With the American economy showing signs of a run down, the Vietnam War unresolved, and the pressing needs for social reform at home, not even the most space-minded members of the American government could justify a headlong plunge into the year 2001.

Thus it was with a comparatively modest $3·7 billion for the year 1970 that NASA drew up its plans for the future. (NASA's 1971 budget was down to $3·3 billion while the US defence budget stood at $73·6 billion.) In contrast the Russian space programme was not showing any signs of a similar slow-down. 1970 saw no decrease in the number of satellites launched. Indeed with an automatic retrieval of a lunar rock sample by Luna 16 in September and the eighteen-day flight of Soyuz 9 with Andrain Nikolayev and Vitaly Sevastiyanov in June, the Soviet space effort was again winning some of the first prizes. This was also in spite of the fact that by late 1970 there was still no sign of the monster launcher, estimated to be one third more powerful than Saturn V, expected by western observers for some time. Luna 16 tended to confirm Russian plans to be content with automated moon exploration for the present, while the Soyuz missions indicated that a Soviet permanently manned space platform is almost certainly the number-one priority. Great emphasis has been given in recent Russian flights to the problem of long periods of weightlessness which must be solved if men are to live and work in orbit for useful lengths of time. Soviet space-medicine experts are not only concerned with the difficulties of adjusting to the zero-gravity state, but also the questions of readaption to earth gravity conditions following the flights.

At the International Astronautical Federation meeting in Konstanz, Germany, in October 1970, Andrian Nikolayev reported that it took him two weeks to recover from the eighteen-day Soyuz 9 flight. His personal view was that some sort of artificial gravity would be necessary if periods longer than one month, one of the known long-term goals of Russian space planners, were to be spent in space. Indications of this Soviet priority came in October 1969 when the Russians launched three manned Soyuz craft in as many days. On 11 October Soyuz 6 with cosmonauts Georgi Shonin and Valeri Kubasov aboard was up in earth orbit. The next day Anatoli Filipchenko, Vladislav Volkov and Victor Gorbatko followed them in Soyuz 7. On 13 October two veterans of the January Soyuz 4 and 5 flights, Vladimir Shatalov and Alexei Yeliseyev, were orbiting in Soyuz 8. The group flight lasted seven days, each spacecraft remaining in orbit for five days. About fifty orbital manoeuvres were carried out both automatically and by manual control. No docking was carried out which led to the speculation that this triple flight was a result of replanning following the loss of the giant booster earlier in the summer. It has been suggested that this was to have launched a major section of an orbiting space station with which one or more Soyuz craft would have docked. The crews also carried out several experiments associated with life aboard a space station including a number of space welding tests. Further indications of the role of the Soyuz craft came in Spring 1970 when the Russians put full-sized models of two Soyuz series on view at Japan's Expo 70 in Osaka. The mock-ups presented Soyuz 4 and 5 as 'the world's first space station'. The linked craft measured 50 feet long with a maximum diameter of 12 feet. Each Soyuz consists of the three modules. A service module houses a restartable engine as well as the main power supply for the spacecraft. In these respects it is similar to the Apollo service module but in addition there are wing-like solar batteries (see diagram on page 36). Forward of the service module is the command module which can accommodate three astronauts. The third module, forward of the command module is used for work and rest in orbit. Housed above this module is the docking tunnel for access between the Soyuz craft. Like the Apollo design this also contains the docking drogue and probe mechanism. For re-entry the crew use the command module, ejecting first the work/rest module, then the service module. Umbilical plug-in units on the exterior of the work/rest module enable cosmonauts to carry out extravehicular work without the spacecraft hatches open, as is necessary in the current Apollo design.

The Americans and Russians are not of course alone. France, Japan and China have each launched their own satellites. The first French orbital launch came on 26 November 1965 when a small satellite was launched from Hammaguir in the now Algerian Sahara. They have since constructed and used their new space centre at Kourou in French Guiana. Japan and China put up their first orbiters in 1970. On 11

February, from Uchinoura in Southern Japan came Osumi, a modest sphere borne aloft by a four-stage Lamda rocket. Then on 24 April China launched a 380-pound satellite, rather larger than expected, which announced its presence with a taped rendition of the song 'The East is Red'. Still waiting on the launch pad in mid 1970 is Britain with its Black Arrow rocket and the as yet ill-fated European cooperative booster, Europa. Other national and multi-national projects exist on paper, both for the construction of satellite launchers and for experimental packages to be flown by nations already possessing the fire power. In 1970, with most national space programmes facing a budget squeeze, there is increased interest in the creation of a global space agency. In particular NASA is making approaches to western nations, the idea being that national projects be supplanted by a stake in a super-NASA.

Moon cities

In 1970 the prospect of a permanent scientific base on the moon seemed rather further off than it did in the opulent days of the late sixties. However NASA has already spent millions of dollars in thousands of different detailed studies to this end. Ultimately it cannot be doubted that a moon city's foundations will be laid. The area of the moon open to exploration is roughly equal to that of Africa and it will provide many years of fruitful exploration. The first moon bases will probably be temporary structures, sufficient for several weeks at a time. Supplies will have to be ferried from earth and the process will be very costly. It is perhaps doubtful that a single nation could foot the bill, unless, and it seems a remote possibility, the moon turns out to have some strategic role for the military. Permanent stations not unlike those in the polar regions of the earth will become more feasible as man learns how to exploit the moon's resources. He will need to tap the sun's abundant energy during the fourteen days of the lunar 'day' in order to survive the equally long lunar 'night'. Protection from the harsh radiation of space, food and water supplies, a breathable atmosphere; all these will have to be produced as much as possible from lunar resources. It will tax the most ingenious engineers, physicists, biochemists and mining-geologists to suggest economically viable answers. Nuclear energy and the artificial synthesis of food-stuffs are already providing clues as to how a lunar colony might survive but in another ten years there will be many radically new ways of solving such problems. At this stage one can only speculate in the manner of the makers of that excellent film *2001*.

Some years ago von Braun stated that he was too old to go to the moon, but that in ten years he might not be. The reduction of space travel to an event no more remarkable than a trip in a commercial jet plane will certainly take more than a decade but that time will eventually come. Perhaps the greatest setback to dreamers of moon cities is the failure as yet to find any water on the moon. Upon the availability of water depend many possibilities. From it oxygen can be extracted for men to breathe and hydrogen to fuel his rockets. Water can support life: not only man's but whatever other earthly organisms he may choose to transplant to the moon. If water can be found buried deep somewhere in the moon, then the prospect of lunar farming becomes a real one and with it a better chance of a reasonable quality of life for man on the moon. Who knows, one day the moon may be the last refuge from a polluted earth. Some airlines have already opened their waiting lists for service which may yet be the only way some of us will be able to visit our grandchildren.

Shuttle services and space stations

By the early 1980s a model of a Saturn V or perhaps a discarded one, will draw the crowds at the Smithsonian Museum in Washington. The primitive monster which carried the first men to the moon will draw a chuckle from young engineers and a nostalgic sigh from their parents. By then NASA's space shuttle will be ferrying astronauts and supplies to and from an earth-orbiting space station. The concept is simple enough. On the launch pad two or more manned vehicles are clustered together. Of these the central one is destined for orbit while the

180

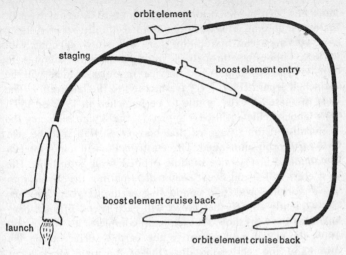

Typical space shuttle concept

other(s) has (have) a supplementary booster role. After staging, each supplementary booster is flown back to an aircraft-like landing-on-wheels while the central element goes into orbit. Its mission complete, the central vehicle re-enters the earth's atmosphere to land like the others. In this way each element of the system can be used over and over again. During the 1960s NASA was building and flying strange craft shaped like half a pear in their efforts to design a craft that could re-enter the atmosphere from orbit and fly down to a conventional wheeled touchdown on a dry land runway. Clusters of vehicles shaped like this will eventually replace the Apollo command and service module in the role of NASA's standard carrier. By the 1980s these shuttles should be flying as many as a hundred trips each year.

Long before the shuttle is flying, NASA will have conducted its first experiments with embryo space stations. Ultimately the plan is to have a permanently manned orbiting platform with fifty to a hundred men on board. Engineers have already converted the third stage of a Saturn V, the Saturn IVB, and fitted it out as a three-man laboratory. The hydrogen fuel tank has been modified to provide about 10,000 cubic feet of habitable volume. This has been divided equally between living and working areas. In addition there will be a 1,000 cubic foot airlock connected through a multiple docking adapter to Apollo-type command and service modules, which will be launched separately to ferry fresh crews to the laboratory. The first of these 'skylabs' is due to be launched in 1972 or 1973. The launch vehicle will be a Saturn V, the skylab replacing the conventional third stage of that booster. Skylab is to be sent into earth orbit unmanned. Three astronauts will join it later in a command and service module orbited by a Saturn IB. The first crew will spend twenty-eight days putting the skylab into good working-order and conducting preliminary experiments. Subsequently two further crews will each spend fifty-six days manning the workshop. As each team completes its work it will leave the skylab and return to the earth's surface using the command and service module. One of the main purposes of the mission from a scientific point of view is to make use of the Apollo telescope mount (ATM). Throughout the Skylab mission this will be used for extensive astronomic observation from a vantage-point free of the filtering and distorting effects of the earth's atmosphere.

Power for the skylab will be obtained both from fuel cells, which will also provide drinking-water for the crew, and from solar panels unfurled in orbit from spines along the converted Saturn IVB. In this respect the skylab will resemble the Soviet Soyuz design. A two-gas atmosphere of oxygen and nitrogen will be breathed. Life will be under zero gravity conditions. Each crew will carry up a sixty-day stock of consumables leaving what it does not use as a safety margin for subsequent occupants. As well as astronomy experiments, medical tests will enable NASA to evaluate the effects of prolonged periods of exposure to space conditions. It is hoped that doctors, physiologists and astronomers will be able to take part directly as crew members. As already pointed out the military also have an interest in the skylab. A number of experiments designed by the US Department of Defence include high resolution photography of the earth's surface.

From what is learned during the missions of one or more

skylabs, NASA will further develop plans for the next stage: a twelve-man space station. Tentative plans for this envisage a forty-foot-long cylinder, thirty-three feet in diameter. It will be launched by a new combination of Saturn stages and unlike the skylab will be extended further by subsequent launches. An attempt to provide artificial gravity is also planned. A spent Saturn stage will be attached to the main body of the space station on a long tether and the whole rotated about the centre of gravity at one end of the station. At four revolutions per minute it is calculated that between a third and three-quarters of earthly gravity could be produced. While this would make a number of things less difficult for scientists aboard the space station, not much is yet known about the associated physiological effects likely to be encountered. Disorientation and dizziness will have to be overcome before a workable system of artificial gravity is established. From this work will come, in turn, the geometry of the 100-man platform, the space base of the future. Sections of this will certainly be provided with per-permanent artificial gravity. Hence a wheel-shape is the most favoured, the rim region having the gravity while the centre, supported by spokes, would be gravity-free. The station's sub-systems: life support, environmental control, communications, attitude stabilization and control and so on, will require a vast amount of power. Fuel cells and solar panels are unlikely to be able to cope with the demand. A thermoelectric nuclear reactor will almost certainly be employed. The space base will take years to build and cost much more than the Apollo moon programme but before 2001 its cluster of lights will be seen in the night sky.

Orbiting eyes

Earth-orbiting satellites have already brought about a revolution in both communications and weather forecasting. They have also reaped a rich reward in new scientific information about the nature of space within our solar system, as well as providing exciting new insights into the organization of the cosmos beyond. Through the 1970s and 1980s the role of the

183

earth satellite and its manned counterpart the space station, will be expanded in both scientific and practical fields. The early 1970s should see a network of reliable navigation satellites as well as the development of a novel form of communications satellite which will be used to beam educational radio and television into village schools in the developing countries. A number of studies have already been carried out in America into the use of this type of satellite to benefit communities in India, Brazil and Africa. Perhaps even more exciting is a broadly based NASA programme in which satellites will be employed to lead to the better husbandry of the resources of our crowded and polluted planet. The earth resources technology satellite (ERTS) will enable man to complete a global inventory of crops, mineral resources, forestry, fresh water supplies, grazing lands, fish shoals, and many other reserves of man's material needs. In addition the earth resources programme calls for provision of better maps, both cartographic and geologic, to help man to locate the raw materials needed for a better quality of life on this planet. The programme will also serve to remind man constantly of the dangers already threatening from the unwise use of these reserves, which have already led to widespread pollution of large areas of the earth as well as its precious atmosphere. Ultimately much of the earth resources work will be carried on within the manned laboratories aboard the orbiting space stations.

Among the planets

Space does not end at the moon. Already a number of unmanned probes have been flown out into the solar system beyond to make contact with our two planetary neighbours. Both America and Russia have a strong programme for the exploration of the planets. Of these, Mars and its two tiny moons, Phobos and Deimos, is first on the list. With a diameter of 4,200 miles, the Red Planet is about half the size of the earth. It revolves on its axis like the earth, a Martian day being thirty-seven minutes longer than ours. The twenty-one photographs of the planet taken by the American probe Mariner 4 in July 1965 show a barren-looking surface not unlike the lunar terrain. The appearance of densely packed craters, similar to the moon's, amazed most of the scientists watching the first frames as they were processed by computers after their arrival from across 134 million miles of space. The craters in these photographs range in size from three to seventy miles across, some with fuzzy edges suggesting that they might be rimmed with frost. The tenuous atmosphere of Mars hardly registers on a barometer but the fact that it exists at all keeps alive the hopes of many biologists still hoping to find some form of living organism outside the earth. Some earthly plants have been experimentally grown for short periods in laboratory conditions simulating the Martian environment. When one remembers that certain species of bacteria, even fungi and algae, can put up with extreme conditions on earth, the possibility of some very primitive Martian life seems quite plausible. We have yet to detect any oxygen in the Martian atmosphere, which appears to consist of carbon dioxide and traces of ammonia and methane.

Life also needs water and on Mars there are seasonally variable white polar caps – much thinner than the ice caps of the Arctic and Antarctic – whose sizes diminish markedly in the Martian spring and early summer, then increase again in the autumn. This change is coupled with an increase in dark patches towards the equator in the Martian summer which could be explained by an annual surge by some primitive plant-like life in these regions. Clouds of what look like ice crystals are sometimes seen high above the red Martian deserts and a very recent optical experiment confirmed the presence of water vapour in the Martian atmosphere, though the amount measured corresponds to a film only five-hundredths of a millimetre thick. The possibility of Martian life remains an exciting enigma.

Mars is on average 142 million miles from the sun, which compares with earth's mean distance of ninety-three million miles. The planet takes about two earth years to complete an orbit of the sun and so varies in its distance from us. Its closest approach is some thirty-four million miles, but when it is on

185

Orbits of the inner planets

Orbits of the outer planets
with Mars orbit reduced to same scale

Sizes and orbits of the planets

the other side of the sun from us it can be further than 200 million miles away. Thus an opportunity to send a spacecraft to Mars occurs only once every two years if the route chosen is to be the shortest one. These 'Mars windows' can be predicted, the next dates being July 1971, August 1973, September 1975, and so on. In fact spacecraft are often sent a few months ahead of these dates so that their arrival at Mars coincides with the dates of nearest approach. In this way the probes' findings are transmitted across the minimum distance. The only successful Mars probe of the first four sent up by both America and Russia, the NASA Mariner 4, was launched on 28 November 1964 and reached Mars on 14 July 1965.

Since it was aimed at a moving target, Mariner 4 was put on a course designed to intercept Mars in its solar orbit and it travelled half way round the sun during its 240-day flight, gradually getting further from earth and nearer to Mars, a journey of 418 million miles. A short cut passing close to the sun and then out again to reach the Martian orbit would have needed a tremendous amount of extra rocket power to resist the sun's gravity.

Interplanetary navigation is thus somewhat like a game of billiards, only the sun is stationary and the movements of the planets and their moons are used to help or hinder the speed of planetary probes. Sending a spacecraft past a heavy planet can considerably increase or decrease its speed relative to the sun.

To send a space probe directly towards the sun it is first necessary to cancel out the earth's orbital speed of 62,750 miles an hour. At this speed travelling away from the earth back along the solar orbit, a space probe would be simply standing still in space relative to the sun and would begin to fall towards it. This is over twice the speed necessary to get as far as the moon and even a Saturn V modified to reach this velocity could not carry a very large scientific package along with it. The problem is solved by employing the billiards technique. By aiming the spacecraft at Jupiter, an enormous planet in an orbit some 364 million miles further away from the sun than we are, it would be possible to use its very large gravitational field to slow the spacecraft down so that it began to fall towards the sun. In this way a Saturn V could carry a very

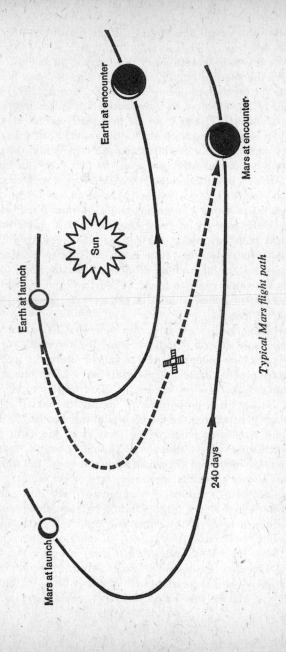

Earth at encounter

Mars at encounter

Sun

Earth at launch

240 days

Mars at launch

Typical Mars flight path

considerable solar probe, though the journey would take much longer.

The technique of interplanetary billiards is already being used to plan the routes of unmanned space probes for grand tours of the solar system in the 1976 to 1980 period. By chance, the alignment of the planets in this period is very favourable. One probe would make a journey out to Jupiter, Saturn and Pluto. Normally a journey to Pluto, which is in orbit some 2,671 million miles away from earth at its nearest approach, would take forty-one years, but by using the gravity-bouncing technique this could be reduced to eight years.

Another such tour has Jupiter, Uranus and Neptune on its itinerary. A journey to Neptune could take up to nine years, even with the bouncing system, but NASA scientists think it is still worth taking advantage of the 1976 to 1980 window.

Before the first grand tour of the outer planets is off the ground, Mars will have received a number of American electronic visitors. Following Mariners 6 and 7, which flew past the planet in July and August 1969, a further two Mariner craft are due to be sent into orbit around Mars in 1971. Then in 1975 the Viking project should see two attempts to land instruments on the planet. The Viking craft will spend ten days in Martian orbit before landing capsules descend to the surface with forty-pound loads of scientific instruments. Landing in the thin Martian atmosphere will present problems and parachutes as well as retro-rockets will be used. Launching the spacecraft in pairs reduces the chances of failure on a mission which can only take place every two years. There are plans to send additional Viking spacecraft in 1977.

'We are much closer to Mars today than we were to the moon eight years ago,' said NASA administrator Thomas Paine recently, giving hope to those who would like to see a manned expedition to Mars in the 1980s. 1985 is a favoured date but it could be sooner.

Problems which will have to be tackled before man can get to Mars are not so much to do with the spacecraft as with man himself. To keep a man alive in space for a trip to the moon and back requires a fair weight of oxygen to breathe and food

189

and water for his nourishment, but for a return trip to Mars, this weight has to be multiplied many times. To get to the planets a system of replenishment of as many of the consumables as possible is needed. On Apollo missions drinking-water is obtained from the fuel cells which use oxygen and hydrogen to generate electrical power for the spaceship. On a Mars mission this water would have to be used over and over again. Its recovery from urine has already been experimentally achieved. To get the oxygen back from the carbon dioxide we breathe out would be another important way of saving weight. An ingenious way of doing this would be to take along some form of rapidly growing edible plant life that does this job on earth and would in addition provide food for the journey. Solid waste could also be used in another system which would employ micro-organisms that would in turn be fed to fish. At first these life support systems might depend solely on non-living technology, but in the future such mixed life environments might prove ultimately more efficient companions for long-distance space travellers.

To keep any life support system going requires power and in the future this will almost certainly be nuclear. The rocket engines that take men to Mars will be developments of nuclear-powered thrusters already being tested on earth. There are other ways of propelling spaceships but nuclear power seems to be the method most likely to follow the present generation of chemical rockets.

The principle of a nuclear rocket engine is very simple. The heat generated by the reactor is used to expand liquid hydrogen which becomes gaseous and is ejected through a nozzle to provide thrust. The difficulties are associated with the very high temperatures that the solid part of the engine would have to withstand.

The psychological problems which will face a group of astronauts within the confined small world of their spaceship also have to be considered. Already both American and Russian volunteers have spent many months as isolated teams. If men can sail alone around the world and spend months down caves without losing their sanity there is no reason why they should not travel to the nearer planets.

The weight and size of a manned spaceship aimed at Mars has been calculated as five times the complete Apollo mooncraft, which means that it could be assembled in earth orbit using five Saturn Vs. Two of them would be used to launch the Mars spacecraft into earth orbit while the other three would put up the elements of the chemical or nuclear rocket with the fuel load of liquid hydrogen. As already mentioned, the hydrogen may even come from the moon or the whole spaceship might be fitted out in lunar orbit. As with the later unmanned probes, the first men will almost certainly go into Martian orbit before sending down a crew in a landing craft similar to the lunar module. Martian escape velocity is 11,400 miles an hour, about twice the lunar escape speed but still half the value for leaving earth, so the rocket to bring them home again would not have to be so powerful.

To get an expedition off Venus, on the other hand, requires an escape velocity of 24,000 miles an hour, just 800 miles an hour less than the earth figure and this is only one of many reasons why Venus, though our nearest planetary neighbour, is not high on our list of cosmic destinations.

Venus, the 'evening star', the 'mystery planet', and home of the Mekon, comes as near as twenty-four million miles on its closest approach. Shrouded in thick cloud, it has defied optical astonomers for centuries. On 12 February 1961 the Russians launched Venus 1 towards the planet, the first attempt to reach one of our neighbours in the solar system. It failed because of faulty communications but later craft, both Russian and American, have flown past, and Soviet capsules have been landed on Venus. All the evidence points to a very hot, turbulent place where man would find it very difficult to survive. The temperature at the surface is well above the boiling point of water and any life at all that might exist would have to lead an airborne existence in the cooler upper layers of the Venusian atmosphere. Venus is not therefore a very enticing place, yet one would like to know more about what goes on under the thick mantle of gases and vapour. Venus is so similar in size to earth and so near to us in the solar system that it is tempting to wonder whether the earth was not once, perhaps only a few score million years ago, in much the same state. Its atmosphere of 98 per

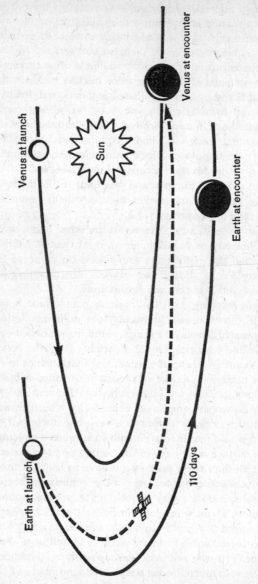

Venus at launch

Sun

Venus at encounter

Earth at launch

Earth at encounter

110 days

Typical Venus flight path

cent carbon dioxide is possibly similar to that of the primitive earth before plant life began to convert it into oxygen. This alone would be a good reason for trying to find out a lot more about our next-door neighbour.

Getting a space probe to Venus involves a similar technique to that used to reach Mars. Firing the vehicle back along the earth's orbital path causes it to lose speed relative to the sun and to fall slowly into the orbit of Venus which is nearer the sun at sixty-seven million miles. The journey takes about three and a half months, under half the time for a trip to Mars. Venus orbits the sun every 225 earth days, which means that a Venus 'window' occurs about every nineteen months.

Between Venus and the sun there is one other planet, the tiny scorched world of Mercury. It is inconceivable at present that anyone would want to go there, but in our geological survey of the solar system it is on the list for an unmanned probe. About the same size as our moon, Mercury travels along an orbit that is within thirty-six million miles of the sun and revolves on its axis very slowly so that a spacecraft would be directed to the terminator, the edge of the planet where the light and dark sides meet. In this place the temperatures might be less hostile to the craft's electronic instruments. The temperature on the dark side of Mercury might, as in the shadows on the moon, be quite low.

Looking out the other way, beyond Mars, a whole host of outer planets remain to be explored. Next to Mars comes the massive planet Jupiter. Its 88,700 mile diameter is ten times that of the earth and only a tenth of the sun's width. Jupiter takes twelve earth years to orbit the sun from which it is on average about 484 million miles distant. Because its gravitational pull is over two and a half times stronger than that of the earth it would be difficult to get a spacecraft off the ground once it had landed, Jupiter has, however, twelve moons. Ganymede, diameter 3,000 miles, is larger than ours. Europa, Io and Callisto have diameters ranging from 1,900 to 2,700 miles and might also be worth investigating as future landing sites.

To earth-bound astronomers Jupiter appears as a stripey planet with a red spot just north of the equator. Its dense atmosphere of hydrogen, ammonia and methane suggests a primeval soup within which the first chemical step towards life may have occurred.

Saturn, nearly as large as Jupiter, comes next at 895 million miles from the sun. Its attraction is due not only to its mysterious rings, but also to its ten moons, the largest of which, Titan, is 3,000 miles across and presents a convenient platform for future space travellers. Saturn's atmosphere is similar to that of Jupiter, and the rings are thought to be fragments of rock, perhaps the result of the disintegration of an erstwhile eleventh moon. Using the principle of interplanetary billiards, nuclear-powered spacecraft could reach Saturn within a reasonable time.

Beyond Saturn is Uranus, which has five moons. It is about half the size of Saturn with a diameter of 30,000 miles. Again a dense atmosphere of hydrogen and methane is indicated. Neptune, another 1,010 million miles out, is in many ways a twin of Uranus. Almost the same size with the same type of atmosphere, it has two moons. One of them, Triton, has a diameter of 2,600 miles.

The last planet we know of in our solar system is Pluto. This remote world is about the size of Mars. It is on average some 3,680 million miles from the sun but its orbit is highly eccentric and from 1969 to 2009 it is in fact nearer in than Neptune. This outpost of our sun has no atmosphere. The next stop in space is another sun called Proxima Centauri or Alpha Centauri C, which is about twenty-five million million miles, 4·3 light years, from our sun.

Outside the Solar System

We do not yet know whether any of these near stars have their own family of planets, but among the 100 billion stars in our galaxy millions of them are similar to our sun. That many of these have planets is statistically probable. That life has, does, or will exist on some of these is also very likely. Professor Philip Morrison, an American cosmologist, has calculated that there are 100 million 'cold' planets suitable for life within our own galaxy and astronomers have already estimated that

194

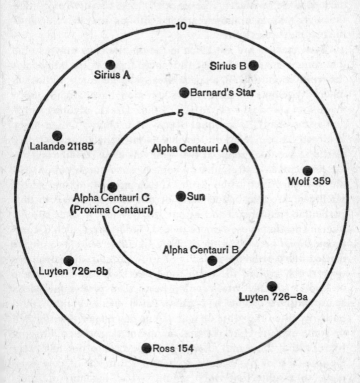

Schematic diagram of the eleven stars within ten light years of the sun

there are 10,000 million galaxies in the observable universe. By a 'cold' planet is meant a body like earth with a temperature range somewhere between that of boiling water and not much below its freezing point. These are the conditions most suitable for the emergence of the large sensitive molecules that over several million years of evolution may develop into the complex we call life.

Given that we are not alone in the universe, we may assume that some beings will be more advanced than we are. Might not the most sophisticated planets already be in touch with each other? Attempts to answer this question have already been made in a sporadic way. In 1960 Frank Drake, a young astronomer at the US National Radio Astronomy Observatory at Green Bank in Western Virginia, pointed the giant electronic 'ear' of a radio telescope at two relatively close sun-like stars in the hope of eavesdropping on some cosmic conversation. No one was surprised at the result. The chances of coming across intelligent life that happened to be directing its signals in that particular frequency band at our solar system at that specific moment in time were remote indeed. To discover such a communications network a much more intensive search would be needed, far beyond the capacity of a few dozen radio telescopes operational around the earth today. Perhaps our best chance of finding another world would be to send probes into deep space with recording and transmission facilities that could patiently listen for years on end. If one day one of them picked up some extra-terrestrial signal an automatic relay could reply by repeating the signal after a given delay in order to attract the attention of the signal's originators. It might of course be naïve to assume that radio would be the medium of intersteller communication. On earth we are already learning to use lasers and who knows what other ways more advanced technologies may have developed.

Getting our space probes out beyond the solar system presents certain problems if the spacecraft is to travel at sufficient speed to get anywhere in the lifetimes of the men who launch it. Theoretically all matter is limited by the speed of light (5,880 billion miles a year). If we could even reach half that velocity a journey to the nearest star would still take over eight years.

Present rocket technology relies on high initial speeds which gradually decrease, whereas to attain half the speed of light, a continuously firing thruster would be needed which would provide steady acceleration. Nuclear rockets would be one solution. Other ideas include using a stream of electrically charged particles, ions, to propel the craft and, more fanciful still, the possibility of constructing giant parabolic mirrors in space off which photons would be bounced. Photons are particles of light and so travel at the speed of light. The principle of the photonic motor is to produce atomic particles and their anti-particles in equal numbers. When a particle and its anti-particle meet they annihilate each other with the emission of a great deal of radiant energy in the form of photons. The idea is to use these to push the probe through interstellar space.

Taking a theoretical photonic spaceship of 200 tons weight, it is estimated that, of this, 150 tons would have to be 'fuel'. In this case it might be possible to reach a speed of 600 million miles an hour after a year of acceleration. This is near the speed of light. Of course, there would be colossal engineering problems. The energy involved would be equivalent to over 1,000 times the amount of electricity produced annually in the world today. No known materials could be used either to handle the particles or for constructing the giant reflector, though magnetic fields could perhaps solve the first problem and some as yet unknown 'superconductor' could be employed to make the mirror.

At present a practical photonic spaceship is unthinkable, but unless we invent some magical 'anti-gravity' device there might come a day in the far distant future when a massive photonic craft weighing thousands of tons and several miles in length might take man on his first trip to the nearest stars, by that time only a few years' journey away.

Priorities

The exploitation and development of space has been the most expensive technological advance man has ever made. Thus the pace with which it will continue must be in keeping with national and international priorities. The speed with which it has been possible to land an American on the moon has undoubtedly been a function of the military background of the missile race and, of course, President Kennedy's decision, in the face of Russian space successes, to turn the moon project into the ultimate symbol of American prestige. Since that decision, the Vietnam war and social problems at home have also become candidates for heavy spending. The enormous cost ($24 billion) has become an issue in American politics though in fact America can afford it. One American writer, Lewis Mumford, has even compared the American space programme with the Egyptian pyramids: 'Both devices for securing, at an extravagant cost, a passage to heaven for the chosen few,' a rather cynical point of view. Even President Kennedy had his doubts and once asked his science adviser: 'Can't you fellows invent some other race, here on earth, that will do some good?' With the pressing needs of the Poverty Programme in America and a hungry world outside, it may seem difficult for some to take the long-term view, yet in the words of Frank Borman returning from man's first trip around the moon: 'Man can now do anything he wants to do technically.' Perhaps he can; whether he wants to remains very much a question of international politics. Yet from space there are no national frontiers and in 1970 came a fresh glimmer of hope for a more united approach to space exploration. In October five NASA scientists visited the USSR Academy of Sciences to discuss compatible docking arrangements for their respective spacecraft. Not only would this make international space rescue a real possibility but it would also create an opportunity for joint experiments in space. Another reason for optimism was a presentation by the Soviet space-medicine chief Professor Oleg Gazenko at the International Astronautical Federation's annual conference in Konstanz, Germany, that same month. He cited the two transatlantic voyages of Thor Heyerdahl's Ra papyrus craft with their multi-national crews. Not only was such international co-operation feasible, but as Dr Yuri Senkevich, the Soviet member of the Ra crew, concluded, it had actually improved as the voyage progressed, especially during moments of crisis. It's all a question, it would seem, of jumping in at the deep end.

Tabular Summary
of Manned Space Flights

Launch date	Astronauts	Spacecraft	No. of orbits	
1961				
12 April	Yuri Gagarin	Vostok 1	1	First man in space
5 May	Alan Shepard	Mercury 3		First American in space, a sub-orbital hop
21 July	Virgil Grissom	Mercury 4		Sub-orbital
6 August	Herman Titov	Vostok 2	17	Twenty-five hours in orbit
1962				
20 February	John Glenn	Mercury 6	3	First American in orbit
24 May	Scott Carpenter	Mercury 7	3	
11 August	Andrian Nikolayev	Vostok 3	64	First double flight,
12 August	Pavel Popovich	Vostok 4	48	nearest approach: three miles
3 October	Walter Schirra	Mercury 8	6	
1963				
15 May	Gordon Cooper	Mercury 9	22	
14 June	Valery Bikovsky	Vostok 5	81	Second double
16 June	Valentina Tereshkova	Vostok 6	48	flight, first woman in space
1964				
12 October	Vladimir Komarov Konstantin Feoktistov Boris Yegorov	Voskhod 1	16	First multi-manned craft
1965				
18 March	Pavel Belyayev Alexei Leonov	Voskhod 2	17	First space-walk by Leonov
23 March	Virgil Grissom John Young	Gemini 3	3	First orbital manoeuvres
3 June	James McDivitt Edward White	Gemini 4	62	First American space-walk by White
21 August	Gordon Cooper Charles Conrad	Gemini 5	120	

Launch date	Astronauts	Spacecraft	No. of orbits	
1965 (continued)				
4 December	Frank Borman James Lovell	Gemini 7	206	First long flight. Rendezvous with Gemini 6.
15 December	Walter Schirra Thomas Stafford	Gemini 6	15	Nearest approach a few feet
1966				
16 March	Neil Armstrong David Scott	Gemini 8	6	First space docking with target vehicle followed by emergency splashdown
3 June	Thomas Stafford Eugene Cernan	Gemini 9	45	
18 July	John Young Michael Collins	Gemini 10	43	
12 September	Charles Conrad Richard Gordon	Gemini 11	44	
11 November	James Lovell Edwin Aldrin	Gemini 12	59	
1967				
23 April	Vladimir Komarov	Soyuz 1	18	Killed when landing parachutes failed, first fatality in space
1968				
11 October	Walter Schirra Don Eisele Walter Cunningham	Apollo 7	163	First Apollo flight following fire in January 1967 which killed Virgil Grissom, Edward White and Roger Chaffee
26 October	George Beregovoy	Soyuz 3	61	Resumption of Soyuz flights, rendezvous with unmanned Soyuz 2
21 December	Frank Borman James Lovell William Anders	Apollo 8		First men to leave earth's gravity, ten orbits of the moon
1969				
14 January	Vladimir Shatalov	Soyuz 4	48	First docking of two manned craft, first exchange of crews
15 January	Boris Volynov Yevgeny Khrunov Alexei Yeliseyev	Soyuz 5	49	
3 March	James McDivitt David Scott Russell Schweickart	Apollo 9	151	First flight of lunar module

Launch date	Astronauts	Spacecraft	
1969 (continued)			
18 May	Thomas Stafford Eugene Cernan John Young	Apollo 10	Lunar module flown to within 30,000 feet of moon's surface
16 July	Neil Armstrong Edwin Aldrin Michael Collins	Apollo 11	First landing on the moon on 20 July
11 October	Georgi Shonin Valeri Kubasov	Soyuz 6	First triple launching. All three craft
12 October	Anatoli Filipchenko Vladislav Volkov Victor Gorbatko	Soyuz 7	completed rendezvous but no docking
13 October	Vladimir Shatalov Alexei Yeliseyev	Soyuz 8	manoeuvre was attempted. Welding experiments conducted by Kubasov.
14 November	Charles Conrad Alan Bean Richard Gordon	Apollo 12	Second moon landing on 19 November in Ocean of Storms.
1970			
11 April	James Lovell Fred Haise John Swigert	Apollo 13	Mission aborted and returned safely to earth following explosion within service module during translunar flight.
1 June	Andrian Nikolayev Vitaly Sevastyianov	Soyuz 9	Longest flight to date (eighteen days). Weightlessness problems investigated.

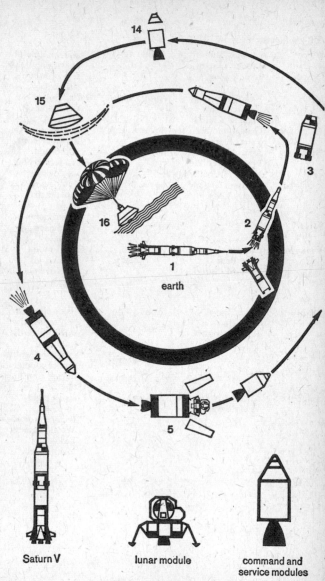

14

15

3

16

2

1

earth

4

5

Saturn V

lunar module

command and
service modules

Apollo 11 mission highlights

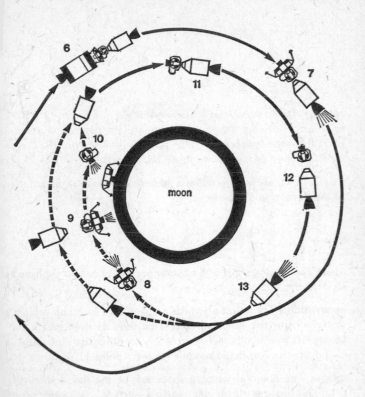

1 launch 2 second-stage ignition 3 third-stage ignition, earth parking orbit
4 translunar injection (TLI) 5 and 6 transposition and docking
7 lunar orbit insertion (LOI) 8 LM–CSM separation
9 LM landing, CSM remains in orbit 10 LM ascends, rendezvous with CSM
11 LM–CSM docking 12 LM jettisoned 13 transearth injection (TEI)
14 CM–SM separation 15 atmospheric entry 16 splashdown

MCC: *'Apollo Nine, this is Houston. We'd like to have a PRD read-out from each of you.'* After a while a puzzled voice replied from the spacecraft: *'Uh . . . roger, thank you. We'll get a PRD report as soon as we figure out what it is.'*

Even astronauts have problems, so here are some explanations of the terms used in this book.

Terms

abort The cutting short of a mission before it has accomplished its objective.

apocynthion The point at which a spacecraft in lunar orbit is farthest from the lunar surface. This term is used when the spacecraft has been launched from a body other than the moon. Example: the command module on the Apollo 11 mission.

apogee The point at which a spacecraft or the moon in earth orbit is farthest from the earth's surface. Example: the Apollo 11 spacecraft in earth parking orbit following launch from Cape Kennedy.

apolune The point at which a spacecraft in lunar orbit is farthest from the lunar surface. This term is used when the spacecraft has been launched from the moon. Example: the ascent stage of the lunar module after it had been launched from the Sea of Tranquillity on the Apollo 11 mission.

attitude The position of a spacecraft in terms of its orientation in space. Since there is no 'up' or 'down', attitude is measured in terms of pitch, roll and yaw (*q.v.*). Example: the attitude of a car turning a sharp banked left-hand corner down a hill has elements of pitch (down), yaw (to the left) and roll (anti-clockwise). For spacecraft, these elements are measured in degrees.

burn-out The point at which combustion ceases in a rocket engine. (The terms shutdown and cut-off are also used.) Example: the burn-out of the third stage of the Apollo 11 Saturn V when it reached the speed necessary for earth parking orbit.

cislunar An adjective referring to the space between the earth and the moon.

de-boost The firing of a rocket engine of a spacecraft in orbit against the direction of flight so that the craft is slowed down and falls into a lower orbit. This is also described as a retrograde manoeuvre. Example: the lowering of the Apollo 11 lunar module's orbit prior to the firing of its descent engine for the landing at Tranquillity base.

delta-V A velocity change.

de-orbit A similar manoeuvre to de-boost but with the intention of taking the spacecraft out of orbit altogether. Example: the descent of the Apollo 11 lunar module to the lunar surface.

down-link The part of a spacecraft-to-ground communication system which handles data originating in the spacecraft. (See also **up-link**.) Example: information on fuel states, cabin temperature and pressure passed automatically to ground stations from the Apollo 11 command module.

entry corridor The critical flight path of a spacecraft before and during re-entry into the earth's atmosphere prior to splashdown.

escape velocity The speed a spacecraft must attain to overcome a gravitational field such as that of the earth (necessary speed: 24,800 miles an hour) or that of the moon (necessary speed: 5,300 miles an hour).

fuel cell An electro-chemical generator in which chemical

205

energy from the reaction of oxygen and hydrogen is converted directly into electricity and water.

G or **G-force** Force exerted upon an object by gravity or by reaction to acceleration or deceleration. We earthlings experience 1 G permanently.

inertial guidance The principle behind the Apollo 11's automatic flight controls. It is a sophisticated navigation system using gyroscopic devices, etc., which absorbs and interprets such information as speed, range attitude, etc., and automatically adjusts the spacecraft systems so that it flies along a pre-selected flight path. Essentially, it knows where it is going and where it is by knowing where it came from and how it got there.

injection or **insertion** A term used to describe the process of putting a spacecraft into a pre-selected flight path or orbit. Example: translunar injection of the Apollo 11 craft towards the moon and lunar orbit insertion once it got there.

penumbra Semi-dark part of the shadow in which light from the sun is only partly cut off on the surface of a planetary body, for example during sunset and sunrise. (See also **umbra**.)

pericynthion The point at which a spacecraft in lunar orbit, launched from a body other than the moon, is nearest the lunar surface.

perigee The point at which a spacecraft or the moon in earth orbit is nearest the earth's surface.

perilune The point at which a spacecraft in lunar orbit, launched from the moon, is nearest the lunar surface.

pitch The movement of a spacecraft about its lateral y-axis. Example: the movement of the prow of a ship at sea.

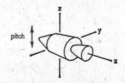

retro-rocket A rocket which gives thrust in a direction opposite to the direction of a spacecraft's movement.

roll The movement of a spacecraft about its longitudinal *x*-axis. Example: a chicken slowly turning as it is roasted on a spit.

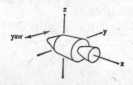

telemetry Measurements taken automatically within a spacecraft in flight and transmitted by radio to the ground. Example: data on astronaut's heart rate, cabin temperature and pressure, etc.

terminator The line on a planet or moon dividing night and day as seen by an observer from space.

translunar, transearth Adjectives describing the voyages to and from the moon respectively.

umbra The dark side of a planet or moon after sunset and before sunrise. (See also **penumbra**.)

up-link The part of the ground-to-spacecraft communication system which handles data originating from the ground. (See also **down-link**.)

ullage A manoeuvre to force weightless propellants in the fuel tanks into the fuel lines. This is usually achieved by the brief firing of a retro-rocket.

yaw Movement of a spacecraft about its vertical *z*-axis. Example: a car turning a corner.

AGS ('aggs')	abort guidance system
AOS	acquisition of signal
ARIA	Apollo range instrumented aircraft
ATM	Apollo telescope mount
BST	British standard time
CAP-COM	capsule communicator
CM	command module
CSM	command service module
DOI	descent orbit insertion
ECS	environment control system
ECU	environment control unit
EMU	extravehicular mobility unit
EPO	earth parking orbit
ERTS	earth resources technology satellite
EVA	extravehicular activity
GET	ground elapsed time
IMU	inertial measurement unit
KSC	Kennedy Space Centre
LCC	launch control centre
LEC	lunar equipment conveyor
LES	launch escape system
LET	launch escape tower
LLRV	lunar landing research vehicle
LLTV	lunar landing training vehicle
LM	lunar module
LOI	lunar orbit insertion
LOR	lunar orbit rendezvous
LOS	loss of signal
LPO	lunar parking orbit
MCC	Mission Control Centre (at Houston)

MCC (followed by a number)	mid-course correction
MIT	Massachusetts Institute of Technology
MSFN	manned spaceflight network
MSOB	manned spaceflight operations building
NASA	National Aeronautics and Space Administration
NASCOM	NASA's communication network
PAO	public affairs officer ('voice' of Mission Control)
PDI	powered descent initiation
PGNCS ('pings')	primary guidance navigation control system
PLSS	portable life support system
PTC	passive thermal control
RTG	radio-isotope thermo-electric generator
SEP	separation
SM	service module
SPS	service propulsion system
S-IC	Saturn IC, first stage of Saturn V
S-II	Saturn II, second stage of Saturn V
S-IVB	Saturn IVB, third stage of Saturn V
TEI	transearth injection
TLC	translunar coast
TLI	translunar injection
VAB	vehicle assembly building
VHF	very high frequency (radio waves)

Index

213

214

More about Penguins
and Pelicans

Penguinews, which appears every month, contains details of all the new books issued by Penguins as they are published. From time to time it is supplemented by *Penguins in Print*, which is a complete list of all books published by Penguins which are in print. (There are well over three thousand of these.)

A specimen copy of *Penguinews* will be sent to you free on request, and you can become a subscriber for the price of the postage. For a year's issues (including the complete lists) please send 20p. if you live in the United Kingdom, or 40p. if you live elsewhere. Just write to Dept EP, Penguin Books Ltd, Harmondsworth, Middlesex, enclosing a cheque or postal order, and your name will be added to the mailing list.

Some other Pelicans are described on the following pages.

Note: *Penguinews* and *Penguins in Print* are not available in the U.S.A. or Canada.

A Star Called the Sun

George Gamow

Sun-worship is perhaps the most rational of all religions:
it pays homage to the source of energy and the most potent
force within certain knowledge. In this century solar
research has advanced at an increasing tempo: today, with
the help of 200-inch telescopes and readings taken outside
the atmosphere, astronomers can precisely measure the
distance, weight, size, composition, temperature, movement,
and life expectation of the sun. This new account of their
findings, by a great American science writer, shows how
research in optics, atomic physics, fluid mechanics, and
other sciences have contributed to calculations so intricate
that only computers can easily handle them.

From the turbulent surface of the sun (chequered with the
spots that influence the growth of plants, radio
communications, and possibly the politics of men) Dr
Gamow takes the reader down into the sun's central inferno
of unimaginable heat. And his chapter on the stellar
population of the universe, with its red giants and white
dwarfs, its pulsating stars and exploding stars, places our
sun in the main sequence of a system in which possibly a
billion other planetary groups exist.

Man and the Cosmos
The Nature of Science Today

Ritchie Calder

That 'science is the everlasting interrogation of Nature by man' is the governing theme of this study by Lord Ritchie-Calder. His aim is to discover 'where science is going', and his survey covers the nature and history of science and the scientific revolution, the dimensions of time and space, and the nature of the universe – both the macrocosmos and the microcosmos. He employs numerous and intriguing pieces of information to illustrate his discussion and shows that it is indeed possible for the layman to gain an insight into the processes of science and its fantastic progression in the second half of the twentieth century.

'. . . knowledgeable and succinct. . . . The book's a splendid read' – *New Scientist*

'This is the best book that Lord Ritchie-Calder has given his readers, and as good as anyone could want for finding out what contemporary science is about' – *The Times Literary Supplement*

For copyright reasons this edition is not for sale in the U.S.A. or Canada

The Nature of the Universe

Fred Hoyle

Moving into the foothills of space travel, we naturally ask if it will really be possible to visit other planets. And can life, as we know it, exist elsewhere in the universe? Or again is Earth, in the face of such monstrosities as the hydrogen bomb, a good risk?

In this revised series of broadcast lectures a Professor of Astronomy at Cambridge gives a lucid outline of our present knowledge about heaven and earth. In doing so he not only answers many anxious questions: he also constructs from observable facts a sober and credible image of an expanding universe which, with its continuous creation of endless wheeling systems, makes the most inspired speculations of poetry and mysticism look pallid.

'No one has expounded the physical cosmos so well since Jeans and Eddington' – *Sunday Times*

'Swift, vigorous and original in matter and in manner' – J. Bronowski in the *Observer*

For copyright reasons this edition is not for sale in the U.S.A.

We Are Not Alone

The Search for Intelligent Life on Other Worlds

Walter Sullivan

The spacial horizon has widened since *We Are Not Alone*
was awarded the Econ Prize for its fascinating presentation
of the greatest question of our time. Walter Sullivan's
informed study of the possibility of intelligent life on other
worlds gains rather than loses in appeal as a result of the
first landing on the moon.

Mathematically the chances of our sharing the universe with
other civilizations – whatever form they may take – are
very high: on the reasonable assumption that such life exists
elsewhere and is technically no less advanced than
ourselves, Walter Sullivan discusses the ways in which
contact might be made and the watches that need to be kept
in order to intercept radio signals from space. For the
emission and interception of signals, within a very narrow
frequency range, offers the best hope of making the most
shattering discovery in the history of man.

Walter Sullivan explores the immense philosophical and
theological problems which must follow such a discovery
and leaves the reader in no doubt of the formidable
difficulties which face interstellar travel when speeds
approaching that of light begin to create the phenomenon
of 'time-dilation'.

'Much the best account of what may happen. He is one of
the very best expositors of science living, and here he is at
his most masterly' – C. P. Snow